PROBABILITY AND STATISTICAL INFERENCE IN ANCIENT AND MEDIEVAL JEWISH LITERATURE

NACHUM L. RABINOVITCH

Probability and Statistical Inference in Ancient and Medieval Jewish Literature

UNIVERSITY OF TORONTO PRESS

Toronto and Buffalo
Printed in Canada
Reprinted in 2018
ISBN 0-8020-1862-9
ISBN 978-1-4875-8524-2 (paper)
LC 79-187394
AMS 1970 Subject classification 01.15

Contents

PREFACE / xi
Guide to Transliterations / xiv
Introduction: Rabbinic Literature / 3

1
DECISIONS AND LOGIC
1.0 Deduction and Induction / 8
1.1 Probability for Decision-Making / 10
1.2 Rabbinic Hermeneutics / 11
1.3 From the Particular to the General / 12
1.4 A Rule for Analogical Inference / 15
1.5 Undecidable Propositions / 20

2
RANDOM MECHANISMS
2.0 Chance and Divination / 21
2.1 Recourse to Chance in the Bible / 22
2.2 Chance Mechanisms in the Talmud / 25
2.3 Lots for the Scapegoat / 26
2.4 Division of the Holy Land / 28
2.5 Redemption of the First Born / 29
2.6 Sharing the Flesh of the Sacrificial Animals / 31
2.7 Assignment of Priestly Duties: The Law of
Large Numbers / 32

3

FOLLOW THE MAJORITY: A FREQUENCY INTERPRETATION

3.0 Acceptance Rules / 36
3.1 Follow the Majority / 38
3.2 Counted and Uncounted Majorities / 38
3.3 Talmudic Acceptance Rules / 39
3.4 Approximate Frequencies / 42
3.5 Standard Majorities / 43
3.6 Half and Half / 44
3.7 Relevant Reference Classes / 44
3.8 A Frequency Interpretation / 47

4

ADDITION AND MULTIPLICATION OF PROBABILITIES

4.0 Arithmetic for Acceptance Rules / 48
4.1 Addition / 48
4.2 Multiplication: Conditional Probabilities / 51
4.3 Multiplication: Independent Probabilities / 54
4.4 Axioms of Probability Theory / 56
4.5 Inverse Probabilities and Bayes's Theorem / 58
4.6 Summary / 61

5

LOGICAL ALTERNATIVES

5.0 The Classical Definition of Probability / 62
5.1 Enumerating the Alternatives / 64
5.2 Combinations *vs* Permutations / 66
5.3 Ultimate Alternatives / 69
5.4 Are Logical Alternatives Equiprobable? / 72
5.5 A Propensity Interpretation of Probability / 74
5.6 What is 'Random'? / 76
5.7 Unknowability and Indeterminacy / 77

6
SAMPLING
6.0 Statistical Inference / 78
6.1 Distribution in a Sample / 80
6.2 Variability in Samples / 82
6.3 From Sample to Population / 84
6.4 Mortality Rates / 86
6.5 Predictive Inference / 88
6.6 Statistical Laws of Nature / 91

7
PARADOXES
7.0 Logical Difficulties / 94
7.1 A Paradox of Indifference / 96
7.2 An Acceptance Paradox / 97
7.3 Bertrand's Paradox / 99
7.4 The Lottery Paradox / 101
7.5 Whence Paradox? / 102
7.6 Shrinking Probabilities / 106

8
EVIDENCE AND ENTAILMENT
8.0 Two Types of Probability / 108
8.1 A Comparative Concept of Evidence / 110
8.2 Legal Presumptions and the Problem of Induction / 112
8.3 A Hierarchy of Presumptions / 114
8.4 Evidence in other Legal Systems / 117

9
INDUCTION AND HYPOTHESIS
9.0 Competing Hypotheses / 118
9.1 An Economy Principle / 119
9.2 Are the Stars Randomly Distributed? / 122
9.3 Falsifying Hypotheses / 125
9.4 A Rabbinic Principle of Maximum Likelihood / 127
9.5 The Modern Principle of Maximum Likelihood / 129
9.6 A Likelihood Ratio / 130

10
SUBJECTIVE PROBABILITIES
10.0 Subjectivistic Probability / 133
10.1 Hunches and Intuition / 135
10.2 Subjective Objectivity / 137
10.3 Degrees of Belief / 139

11
COMBINATIONS AND PERMUTATIONS
11.0 The Pure Mathematics of Probability / 142
11.1 Letters and Words / 143
11.2 Conjunctions of Planets / 145
11.3 A Major Branch of Mathematics / 147
11.4 An Application to Probability / 148
11.5 Combinations in Gaming / 151

12
THE HISTORICAL PERSPECTIVE
12.0 Demonstration and Dialectics / 152
12.1 Etymology and Meaning / 154
12.2 Equiprobability / 157
12.3 Greek Combinatorial Analysis / 160
12.4 Inheritance Problems: Mathematical Expectation / 161
12.5 Legacies, Wagers, and Probability / 165
12.6 Fourteenth-Century Probabilism / 166

13
CONCLUSIONS
13.0 What is 'Rational'? / 170
13.1 The Antiquity of Probability and Decision Theory / 172

APPENDIXES
A Glossary of Hebrew Terms / 176
B Glossary of Mathematical and Technical Terms / 178
C List of Symbols / 180
D List of Tractates in the Mishnah / 182
E Talmudic Teachers with Dates / 183

BIBLIOGRAPHY
Rabbinic Works / 185
General Works / 190

INDEX / 201

Preface

In recent years, historians of mathematics and philosophy have traced the beginnings of probability theory far beyond Pascal and Fermat who long were thought to have invented it *de novo*.[1] These efforts have led to investigation of ancient gaming practices, as well as to a careful study of some aspects of medieval epistemology and logic.[2] Although the Jewish scientific tradition was of major importance in the Middle Ages, very little has been done to explore the history of statistical and probabilistic reasoning in Hebrew literature. A contributing factor is no doubt the erroneous persistent impression that the origins of probability are somehow exclusively associated with gambling problems. Since biblical and rabbinic Judaism was uncompromisingly hostile to these pursuits, even historians of Jewish mathematics were misled to the point that they overlooked a very important feature of rabbinic thought. Thus, as recently as 1962, the article on probability in the *Encyclopaedia Hebraica* asserted unequivocally that 'ancient Jewish thought did not know the concept of probability ... It seems that the very con-

1 'Neglecting the trifling hints which may be found in preceding writers, we may say that the Theory of Probability really commenced with Pascal and Fermat.' Isaac Todhunter, *History of the Mathematical Theory of Probability* (Cambridge, 1865), p. 20

2 Some of these works are listed in the Bibliography

cept of probability is foreign to Judaism.'[3] When the first paper[4] on this subject appeared – and that as recently as 1966 – it was devoted entirely to procedures such as lots and other random devices, as if gaming were the only context in which probability belongs.

In fact, a survey of probability in the ancient and medieval Hebrew sources reveals not inconsiderable attainments in applications of statistical inference and probability logic to a wide variety of legal problems. It will be shown that there was an awareness among the rabbis of different conceptions of probability.

Thus there is a frequency interpretation according to which, associated with the event of drawing an unknown object from a mixture of n As and m Bs, there is the probability $p(A)$ that the object drawn is an A and this is given by the ratio of the number of As to the total, namely $p(A) = n/(n + m)$. On this view an inference is permissible only if the relative proportions of As and Bs can be empirically determined.

Another approach applies to certain types of individual events. If the alternative possibilities can be enumerated they can be regarded as equally probable; a numerical probability can thus be assigned to each alternative.

Still another view regards evidence as endowing relevant propositions with a measure of probability, thus rendering certain conclusions rationally acceptable to a greater or lesser degree.

Moreover, the rabbis formulated the rudimentary rules of the arithmetic of probabilities. They also discussed a variety of theoretical questions about the scope of applicability of the different concepts and the logical and philosophical difficulties to which these give rise, and they suggested decision-theoretic guidelines for accepting and rejecting competing hypotheses.

This is all the more remarkable since there seem to be no known parallels in Greek or other writings of antiquity, although the re-

3 *Encyclopaedia Hebraica*, vol. 14 (Tel Aviv, 1962), 920–1

4 A. M. Hasofer, 'Random mechanisms in Talmudic literature,' *Biometrika*, 54 (1966), 316–21

sults of the present enquiry will, I hope, suggest the need for a careful scrutiny of ancient sources beyond the limits of purely mathematical texts, especially in the Babylonian tradition, the medieval Arabic, and that of the Scholastics. While this work is devoted primarily to presenting and analysing materials from Jewish authors, some attempt has been made to indicate points of possible contact with other cultures, albeit rather sketchily.

The chapter Random Mechanisms has been placed before the other more comprehensive developments, mainly because it is almost self-contained and because it refers back to the very earliest biblical sources. The chapter Combinations and Permutations treats a subject that has always been intimately connected with probability, although the stimuli for its development were quite varied. Because of the evidences of intercultural influences, this chapter seems to go together with that called Historical Perspective.

It is a pleasant duty to record my debt of gratitude to Professor Kenneth O. May who followed the manuscript from its earliest draft through to its final form. The work has benefited from his comments and suggestions in ways too numerous to mention.

I am grateful to my wife, Ruth, without whose constant encouragement this book would never have been undertaken or completed; to my daughter, Tovit, who typed the early drafts and Miss Ruby Ezekiel who did the final version; and to Mrs Gertrude Stevenson of the Editorial Department of the University of Toronto Press, whose copy editing and careful attention to detail saw the book through the press.

The publication of this book has been made possible by the aid of a grant from the Humanities Research Council, using funds provided by the Canada Council.

N.L.R.

GUIDE TO TRANSLITERATIONS

ח = ḥ = ch in Scottish *loch*
י = J, Y = y in yonder
כ = kh

Names which are very common in English, e.g., Jonah, Joseph are given in their usual spelling rather than phonetically: Yonah, Yosef.

PROBABILITY AND STATISTICAL INFERENCE IN ANCIENT AND MEDIEVAL JEWISH LITERATURE

Which is a good way that a man should cleave to? Rabbi Shimon says: 'He who sees what will come to be.' That is, he who learns from the present for the future.
In the sciences, it is an intellectual virtue and it means to infer from the seen to the unseen. However, the intent here is to see to the affairs of man upon which his existence depends – that he should see the consequences of his deeds ... that is a moral quality.

Moses Maimonides
Commentary on the Mishnah: Avot II 12

Introduction: rabbinic literature

0.0 Most of the extant works of Hebrew antiquity are, like the Bible, compilations spanning many centuries. The sages who lived and taught during the first two centuries of the Common Era (approx.) are known as Tanna'im (singular: Tanna), and they are usually designated by generations. The teachings of the Tanna'im and their predecessors are collected in (a) works like Mekhilta, Sifra, and Sifri which are arranged as a running commentary mainly on the legal matter in the biblical books Exodus, Leviticus, and Numbers-Deuteronomy, respectively, and (b) in the Tosefta and similar codices which are divided into tractates according to subject matter. The Tannaitic period came to a close with the redaction of the Mishnah by Rabbi Judah the Prince (born C.E. 135) near the end of the second century. The Mishnah is divided into sixty-three tractates and became the major subject of study in the Academies of the Amora'im (singular: Amora), as the sages were then called. The discussions of the Amora'im ranged over the whole compass of the traditional lore and are called Gemara. Talmud refers to both Mishnah and Gemara together.

There are two Talmuds. The Jerusalem Talmud was brought to its essentially present form by the end of the fourth century and the more voluminous Babylonian Talmud about a century later.

0.1 In the period before the destruction of the Second Temple

(C.E. 70) the Jewish settlement in Babylon began to grow in numbers and in intellectual stature, a development which was enhanced as the heavy hand of Rome made life increasingly difficult in the land of Israel. The Babylonian Gemara is mainly the product of the flourishing schools which thrived within the autonomous Jewish community that lived under the usually benevolent protection of the Persian empire.

In Israel meanwhile conditions continued to worsen. After centuries marked by bloody convulsions of rebellions followed by mass slaughter and exile into slavery, the adoption of Christianity by the Emperor Constantine as the state religion led to aggravated persecution, including the forced closing of the academies and the banishment of the teachers of the Law. Thus the Jerusalem Talmud was never really brought to completion and was greatly overshadowed in subsequent times by the more comprehensive Babylonian Talmud.

0.2 For many centuries, throughout the Diaspora, Bible and Talmud were the major subjects of study for Jews. Since study was regarded as a primary religious obligation, most of the best minds devoted themselves assiduously to rabbinic learning, producing a vast library of Talmud-based literature. This includes commentaries as well as self-contained treatises, codes of law, and commentaries on commentaries. Of great importance too, are the many volumes of responsa in which distinguished authors answered queries on Jewish law and lore. Of those rabbis whose opinions are discussed in this study, a few will be mentioned here to give some idea of the geographical distribution. The dates of all authors cited appear in the bibliography.

One of the earliest expositors to write a running commentary on the Babylonian Talmud was Rabbi Gershom ben Judah (960–1040). Born in Metz, he made immense contributions to the organization of Jewish life in France and Germany and to the spread of Torah learning there. The commentary *par excellence* on the Babylonian Talmud was written by Rabbi Shlomo Yitzḥaki (1040–1105) of Troyes in France, who is known by the acronymic RASHI. Without Rashi's explanations, the Talmud would be almost a sealed book.

Out of Rashi's school came several generations (thirteenth century) of scholars who composed glosses on the entire Babylonian Talmud called Tosafot (= addenda).

Rabbi Abraham ben Meir Ibn Ezra (1092?–1167?) was driven by oppression and poverty from Spain and France to Egypt and North Africa, and as far as England. Although he knew hardly a moment's peace, he was a highly creative writer whose contributions to many areas of science and learning have earned him lasting acclaim.

Also born in Spain, Rabbi Mosheh ben Maimon (1135–1204), known in English as Moses Maimonides was forced by religious persecution to flee and lived most of his life in Egypt. In addition to a commentary on the Mishnah, he produced a systematic codification in fourteen volumes of all the legal material in both Talmuds, as well as all other ancient canonical sources. He is also the author of one of the most important philosophical works of the Middle Ages, *The Guide of the Perplexed*, which like his commentary on the Mishnah was written in Arabic.

In Spain, Rabbi Mosheh ben Naḥman (1194–1270?), Rabbi Shlomo (ben Abraham) ben Adret (1235–1310), and his disciple Rabbi Yom Tov ben Abraham of Seville (*fl.* early fourteenth century) wrote Talmud commentaries as well as independent works dealing with legal and religious questions.

Meanwhile, Jewish scholarship blossomed in France with men such as Rabbi Abraham ben David (1125–1198) of Posquieres, Rabbi Menaḥem ben Shlomo Meiri (1249–1319?) of Perpignan, and Rabbi Levi ben Gershon (1288–1344?), who was a distinguished astronomer and mathematician as well.

From Germany, we have a code by Rabbi Mordecai ben Hillel Ashkenazi who, with his wife and five sons, was martyred in Nuremberg in 1298. Another major codifier, Rabbi Asher ben Yeḥiel (1250?–1328), was born in Germany but was compelled to wander to Spain in his later years. His code together with that of Maimonides became the foundation for all subsequent development.

Their responsa alone assure leading roles in cultural history for Rabbi Isaac bar Sheshet who was born in Valencia (1326) and died in Algiers (1408), and for Rabbi Joseph Colon (1420?–1480) who lived most of his life in Italy. A classic treatise on Talmudic logic was written in 1467 by Rabbi Yeshuah Halevi ben Joseph (1440–1500) of North Africa.

While major interest in this work centres on Talmudic concepts, the views of the medieval rabbis are important in their own right, not only as exposition of the Talmud. For this reason, I have referred to writers up to and including the sixteenth century. Among these are the great codifier Rabbi Joseph ben Ephraim Caro (1488–1575), whose family was exiled during the expulsion of the Jews from Spain (1492) when he was in his fifth year. After many wanderings, he finally reached the land of Israel. His prolific output includes a commentary on Maimonides' Code. From the Holy Land came also Rabbi Bezalel ben Abraham Ashkenazi (1520?–1589) who wrote a compendium and digest of early commentaries and annotations to the Talmud. In Poland, Talmudic learning was immeasurably deepened by Rabbi Shlomo Ben Yeḥiel Luria (1510–1573). In the second half of the century there flourished in Greece Rabbi Joseph di Trani the Elder who is known for his responsa and novellae on the Talmud.

Most of the commentaries mentioned dealt mainly with the Babylonian Talmud and only occasionally or indirectly with the Jerusalem Talmud. It was not until the latter half of the eighteenth century that extensive commentaries were written on the Jerusalem Talmud.

For a detailed introduction to the Talmudic literature, see Herman L. Strack, *Introduction to the Talmud and Midrash* (Philadelphia, 1945).

0.3 The works of most of the writers cited have not been translated into English. However, where I know of the existence of such translations, these are indicated in the Bibliography. For Bible passages, I have used the Revised Standard Version, and for Maimonides' *Guide of the Perplexed* the new translation by Pines. In all other instances, the translation is my own. Although the Babylonian Talmud and much of Maimonides' Mishneh Torah have been rendered into English, unfortunately the different volumes are from different hands, and thus the same technical terms appear in different English equivalents in different places. Moreover, in a study of this kind, I feel it is necessary to convey the literal sense of the quoted texts as accurately as possible without regard to literary fancy. I hope that the sacrifice of stylistic elegance will be justified by added precision in conveying the views of the ancient writers.

For non-Hebrew writings, the translation used is given in the Bibliography. Where only an edition in the original tongue is listed, the translation is my own.

Although both Talmuds are arranged like the Mishnah, the standard form of references to the Babylonian Talmud is to tractate, folio number, and side, while for the Mishnah, Tosefta, and the Jerusalem Talmud it is tractate, chapter, and section. To avoid confusion, such references always specify before the tractate – M = Mishnah, T = Tosefta, or J the Jerusalem Talmud. The commentaries and novellae on the Talmud are referred to by indicating the Talmudic passages they deal with. For the codes, the form of references is the title of book, followed by number of chapter and paragraph, while for the responsa the number of the responsum is given. Where a specific passage may be difficult to locate in a lengthy responsum, I have given a page or folio and column reference to the edition I used which is listed in the Bibliography.

1
Decisions and logic

1.0 DEDUCTION AND INDUCTION

All rational thought falls into two main forms – deduction and induction. Deductive reasoning proceeds from premises and reveals consequences that are implied by them. If the premises are true, the conclusions deduced from them in accordance with the rules of deductive logic must be true as well. Where are the premises to come from? They themselves may be derived by deduction from other premises, but if there is not to be infinite regress, one must ultimately reach prior premises or axioms which are assumed to be true but which are not deducible from others. Nothing more can be demonstrated deductively than is already contained implicitly in these axioms. Thus, if deduction is to yield useful and significant conclusions, there must be more or less general propositions posited as premises.

Now observation can yield only particular data, not general propositions. For example, from the fact that all the swans we have observed in the past have been white we cannot deduce that all swans that may ever be observed will necessarily be white. Nor is it a contradiction to assert that even the very next swan to be observed will be black.[1] Does this mean then, that all our past

1 This argument goes back to David Hume: ...'even after the observation of the frequent or constant conjunction of objects, we have no reason to draw any

observations can tell us nothing at all about the unobserved? To add to our store of knowledge, to move from known premises to consequences that go beyond them, inductive inference is required. Inductive thinking has been called ampliative[2] because it alone amplifies the scope of knowing. Induction is really a form of conjecture in which on the basis of limited available evidence one projects beyond the known facts to what is rationally acceptable or credible about the unknown. As such, its conclusions are not unfailingly true. They must always be regarded as provisional and subject to modification in the light of further experience. However, since the human condition precludes possessing absolute knowledge of the requisite universal premises for making valid deductions in most contexts of interest or importance, principles which enable us to make reasonable inferences are the key to all scientific progress. For, as Lord Keynes observed, 'in the actual exercise of reason we do not wait on certainty.'[3]

Inductive thinking leads to what are known as probable conclusions, in the sense that there is a variable element of uncertainty in them. A prominent feature of inductive procedure is the use of statistical data and statistical inference. These have grown steadily in importance since the rise of empirical science in the seventeenth century. The developments in physics as well as in the mathematical theory of probability in this century have changed our conceptions to such an extent that we no longer expect the laws of nature to be construed absolutely. Rather, statistical laws and statistical inference have become the distinguishing feature of science. Probabilistic reasoning is now central in all rational thought.

In view of the pervasive nature of probability-type logic which permeates all realms of life and civilization, it would indeed be very strange if it were entirely the product of modern times. To appreciate its importance in ancient and medieval Jewish culture, it is

inference concerning any object beyond those of which we have had experience.' *Treatise of Human Nature* Bk. I, chap. 3, sec. 12, ed. T.H. Green and T.H. Grose (London, 1874), 436. See also J.S. Mill, *A System of Logic*, 8th ed. (New York, 1888), Bk. III, chap. 3, sec. 2, 226

2 See, for example, Charles S. Peirce *Collected Papers*, ed. Charles Hartshorne and Paul Weiss (Harvard University Press, 1931–5), 2.619, 2.709

3 J.M. Keynes, *A Treatise on Probability*, 3rd ed. (Oxford, 1961), p.3

necessary to inquire into some basic aspects of Jewish intellectual history.

1.1 PROBABILITY FOR DECISION-MAKING

Any attempt to understand why and how probabilistic reasoning arose among the Jews, must be based upon an insight into the rabbinic conception of the Law.

The Law, though originating in Revelation, is viewed by tradition as an eminently rational pursuit. Moreover, Jewish law is not only or even primarily remedial law concerned with adjudicating quarrels and punishing wrongdoing. It deals mainly with social, ethical, and ritual duties, and it operates under the assumption that every situation in which a man finds himself has implications which need to be spelled out in terms of responsibilities and obligations as well as rights and privileges. While the ancient rabbis naturally accepted the possibility of special and explicit Divine guidance in specific circumstances, they insisted unequivocally upon the efficacy of rational methods to ascertain what course of action the Law demands.[4] Furthermore, except on the rarest occasions, they refused to grant even apparently miraculous inspiration the authority to override the directives of the Law as developed through application of the accepted logical and procedural rules.[5]

Since almost every thing has legal consequences, the need was obvious for conceptual tools to cope with the host of uncertainties that arise in daily life; to weigh inadequate evidence and to determine criteria of degrees of certainty. When not all the facts are known or even attainable, duties are not suspended. Rather judgment is required to decide how to act in the light of what information is available. Since divination is effectively precluded for religious reasons,[6] criteria for deciding in doubtful cases do not reveal by supernatural means that which is hidden. Rather they spell out rationally acceptable standards for acting on certain assumptions.

4 Thus the Talmud states: 'It is written, "These are the commandments" (Leviticus 27:34) – no prophet may innovate anything from now on' (*Shabbat* 104a)

5 See *Bava Batra* 12a: 'A sage is superior to a prophet.'

6 See below p.22

The matter is perhaps best expressed in the words of the thirteenth century sage Rabbi Asher ben Yeḥiel.

> We have been commanded to pronounce a true verdict. Even though we are not prophets to judge matters concealed in the heart, yet 'a sage is superior to a prophet.'[7] We follow in the footsteps of the ancients and learn from their deeds ... In many cases the Talmudic sages decided in accordance with evident probabilities.[8]

With this approach, it is natural that probability reasoning appeared early in the Jewish schools, and was applied to almost every area of the law, as a technique for deciding between alternative courses of action.

1.2 RABBINIC HERMENEUTICS

Coupled with this practical decision-theoretic approach to the evaluation of evidence was a highly specialized logic of induction which answered the needs of interpreting the biblical text. Most of the legal prescriptions in Scripture are in the form of case law, rather than general principles. In order to extend the law to all relevant instances, general concepts must first be inferred from the sacred text. Thus, hand in hand with the 'Written Teaching' – as the Scriptures are known among the Jews – went the 'Oral Teaching,' or the interpretation which enabled one to derive guidance for novel situations. Given general premises, deductive logic can explore all their ramifications, but to formulate universals on the basis of particulars requires a different kind of reasoning. Furthermore, the canons of interpretation must satisfy certain specific requirements rooted in the nature of the particular text in addition to expressing universal aspects of rational thought. While, therefore, the norms of deductive inference are taken for granted in the rabbinic sources, precise rules are laid down for interpretation. These are known as the 'thirteen rules by which the Torah is expounded,' and in the formulation of the martyr Rabbi Ishmael

7 *Bava Batra* 12a
8 Rabbi Asher ben Yeḥiel, *Responsa* 68:23 (p.126, col.1)

(early second century) they appear in the introduction to Sifra, which is the Tannaitic exegesis of Leviticus. The rules themselves go back much earlier, but Rabbi Ishmael's version has become best known since it was incorporated into the daily service as part of the required Torah study incumbent upon all Jews. As such it appears in every prayer book.

It would take us far beyond the concerns of the present inquiry to examine the 'thirteen rules' in detail. However, some insight into their inductive character will help to throw some light on the origins of statistical inference. To this end several simple examples will be cited.

1.3 FROM THE PARTICULAR TO THE GENERAL

Rule 3 is entitled *Binyan Av*, literally 'building a father,' where father is used in the sense of a general principle which stands in relation to the particulars as a father to sons. A *binyan av* allows the derivation of a general law from one or more specific instances. It is postulated that there are no superflous elements in Scripture. If another case is mentioned, it must be because an analogy derived from one case alone will not do. Thus all scriptural statements are regarded as independent axioms and this independence is to be proved.

A somewhat similar situation obtains in modern mathematics, where, in the axiomatization of a theory, one seeks the smallest set of axioms from which the theory as a whole can be deduced. To this end, it is necessary to prove that each axiom is independent of the rest, namely, that it is not implied by them. In the Talmud, however, the problem is not to reduce a given axiom-set to the smallest equivalent set. For, it is assumed that the text of the Pentateuch is irreducible, and it is an exercise in logic to demonstrate it.

In our example, the rule of *binyan av* is applied to the law of torts. In the Bible several cases of damages are mentioned. 'When an ox gores ... ,'[9] 'When a man digs a pit and does not cover it

9 Exodus 21:28

... ,'[10] 'When a man causes a field or vineyard to be grazed over ... , When a fire breaks out and catches in thorns ... .'[11] In all these events, the owner or originator of the property that causes the damage is held responsible and must make restitution.

What is the general criterion that determines liability? Each individual case differs from the others in some ways. Yet, if they were completely different from each other in all respects save that there is liability to pay for damage caused, one would be forced to conclude that the law applies in these cases alone and no others. However, if one can determine one or more characteristics which all four have in common besides that of liability incurred for damages, the rule of *binyan av* provides that the largest set of common characters somehow forms an inseparable unit, and every case exhibiting all the remaining features must also be subject to the law of liability, even though Scripture does not explicitly mention it.

The rule of *binyan av* is equivalent to a stronger form of what is known in general logic as analogical inference,[12] defined as follows: given that each member of a set S has the properties that define the class K, as well as the further property P, and given also that x resembles the members of S in being a member of K (i.e., having the properties that define K), conclude that x also resembles the members of S in having P.

In the case of the *binyan av* in our example, S is the set of the four torts mentioned in the Bible, P is the property of liability, and the properties which determine K must be discovered by careful analysis.

The Mishnah says:

The four principal causes of damage are the ox, and the pit, and the tooth [grazing], and the outbreak of fire.

[The case of] the ox is not like [that of] the tooth and [the case of] the tooth is not like [that of] the ox; nor are [the cases of] both, which are living things, like [that of] fire which is inanimate; nor are [the cases of] these,

10 Ibid. 21:33

11 Ibid. 22:5, 6

12 See Arthur Pap, *An Introduction to the Philosophy of Science* (1962), p.146

whose manner is to go forth and do damage, like that of the pit which is stationary.

The common feature in all these cases is their aptness to do injury, and the care of them devolves upon you, and if one of them caused damage, the one responsible for the damage must pay compensation with the choicest of his land.[13]

The reasoning is straightforward. The assumption is that there is a uniform association of characteristics in all cases. Thus, one first eliminates the differentia of the four types of torts. The common features are then found to be three: (1) an object that potentially causes damage, (2) someone is charged with responsibility to care for the avoidance of damage, (3) that person is obligated to pay compensation if damage occurs. The general principle follows directly.

In the Talmud, such reasoning can be demolished if it can be shown that the cluster of common features is not maximal, or that in other instances the sought-for feature belongs to a different cluster of common ones. Before a *binyan av* is admitted as valid, attempts are made to undermine it in this way. Only if it can survive all such attempts is it finally accepted as true.

The first of Mill's celebrated canons of inductive reasoning is named by him 'the method of agreement.' It states: 'If two or more instances of the phenomenon under investigation have only one circumstance in common, the circumstance in which alone all the instances agree, is the cause of the given phenomenon.'[14] Mill attributes a single effect to a single cause: at least whenever he can. The Talmudic criterion is more general – the maximal set of circumstances in which all the instances agree must be seen as the 'cause' of the given phenomenon.

In accordance with the principle that Scripture does not mention a particular case if it can be derived by analogy or induction from others already stated, the *Amora-im* analysed the *binyan av*

13 *M. Bava Kama* I 1–2. See Rabbi Yeshuah Halevi ben Joseph, *Halikhot Olam*, IV 2 for a complete treatment of *binyan av*

14 J.S. Mill, *A System of Logic*, Bk. III, chap. 8, p.280

of torts to ascertain whether it could have been established on the basis of any three only of the four given instances.[15] For example, Rava shows that 'goring ox' cannot be derived from the other three. For, they are all initially likely to cause damage, whereas ordinarily a tame ox is not likely to gore. If only the three instances were named, one might suppose that the owner's obligation to take precautions, as enjoined by Scripture, extends only to likely situations. By mentioning the case of 'ox' as well, Scripture precludes this presumption.

The complete argument to show that each case is essential need not detain us here. For our purposes it suffices to see that *binyan av* is a specialized inductive inference – a carefully defined analogy.

1.4 A RULE FOR ANALOGICAL INFERENCE

An even more interesting kind of hermeneutical argument closely related and logically prior to *binyan av* is rule one entitled *Kal Va-ḥomer*. The literal meaning of *kal va-ḥomer* is 'lightness and severity.' This terminology can be understood in the light of Aristotle's definition of a syllogism in Barbara which is a special case of *kal va-ḥomer*: 'Whenever three terms are so related to one another that the last is contained in the middle as in a whole, and the middle is ... contained in ... the first as in ... a whole, the extremes must be related by a perfect syllogism.'[16] Thus

Every C is B,
Every B is A,
Therefore,
Every C is A.

Such an argument depends for its validity upon the existence of a relationship of set inclusion; for simplicity we take it to be proper inclusion. (Every C is B, but not every B is C, etc.) This determines an ordering of the sets. Thus in set-theoretic notation the syllogism

15 *Bava Kama* 5b
16 *Analytica Priora*, Bk. I, chap.4, 25b 32

can be rewritten $(C \subset B) \& (B \subset A) \rightarrow (C \subset A)$: read, if C is contained in B and B is contained in A, then C is contained in A. The ordering thus determined is $A \supset B \supset C$. Aristotle's 'middle' and 'extremes' refer to the terms. The Talmudic 'lightness' and 'severity' convey the sense of ordering.

Now a *kal va-ḥomer* extends the concept of ordering relation beyond that of set inclusion. It is postulated that any two elements are comparable, though the relation need not be transitive. The ordering is defined as follows: given a set of subjects $x, y, z, \ldots$ and a set of attributes $A, B, C, \ldots$, which can be predicated of the given subjects, we write $A(x)$ to represent the proposition 'x has property A,' and $\neg A(x)$ means 'not-$A(x)$' or 'x has not property A.' If now it is known that $A(x)$ but $\neg A(y)$, then we say x is more weighted or more severe than y, or y precedes x.

$A(x) \& \neg A(y) \rightarrow y < x$, read: if x has property A and y has not property A, then y precedes x. (The symbol $<$ represents the ordering relation.) If it is also given that $B(y)$, the rule of *kal va-ḥomer* postulates that $B(x)$ also. The rule says: what is true of the lighter is also true of the more severe.[17] We can write the rule in symbols as an axiom-schema:

$$y < x \& B(y) \rightarrow B(x),$$

read: if y precedes x, and y has property B, then x has property B. It is clear that this is very far removed from a syllogism. In fact without further qualification the rule might be used to infer almost anything. There are two restrictions on the applicability of the *kal va-ḥomer* rule:

R1 If it can be shown that there exists an attribute C, such that not-$C(x)$ but $C(y)$. This implies the reverse ordering $x < y$. $C(y) \& \neg C(x) \rightarrow x < y$.

When both $x < y$ and $y < x$, the rule of *kal va-ḥomer* does not apply. The significance of this restriction will be discussed below.

17 A special case of *kal va-ḥomer* is what Aristotle calls a fortiori. 'The argument that a man who strikes his father also strikes his neighbours follows from the principle that, if the less likely thing is true, the more likely thing is true also; for a man is less likely to strike his father than to strike his neighbours' (*Rhetorica* II 23, 4; 1397b15.) The novel element in *kal va-ḥomer* is the purely formal definition of ordering.

R2 If there exists a subject z such that A is predicated of z and not-B as well. $[A(z) \,\&\, \neg B(z)]$.

Since $A(z) \,\&\, \neg A(y) \rightarrow y < z$, if the inference by *kal va-ḥomer* were valid, one could then conclude $y < z \,\&\, B(y) \rightarrow B(z)$, which is known to be false. Therefore the rule does not apply in such a case.[18]

What is accomplished by the *kal va-ḥomer* is a special kind of analogy. For, in order to apply the rule to two subjects x and y, we must be given two premises: (1) $A(x) \,\&\, \neg A(y)$ to establish $y < x$, and also (2) $B(y)$. Now, in general, y will have other attributes also, say $C(y)$, $D(y)$, etc. For the rule to be applicable, in the worst case, it is not known whether $C(x)$, $D(x)$, etc. It certainly cannot be given that $\neg C(x)$ or $\neg D(x)$ etc., for in that case restriction R1 would bar the inference. Now by the rule of *kal va-ḥomer*, x is assumed to be analogous to y in being endowed also with those predicates about which we have no given information whether or not they apply to x. The cluster of common predicates thus defined is minimal in the sense that wherever we find one, the others can co-exist; for, if not, restriction R2 would apply. Thus, there is no inconsistency in extending the analogy to all the relevant characters.

Let all the predicates A_i such that $A_i(x) \,\&\, \neg A_i(y)$ be $A_1, A_2, \ldots,$ A_m, and let $B_1, B_2, \ldots, B_n$ be the attributes predicated of y. If some of the B_i are not given to be true of x, then the *kal va-ḥomer* postulates we can infer that to be the case. Thus $B_i(x) \,\&\, B_i(y)$ for all $i = 1, 2, \ldots,$ n. Moreover, there exist no C_k such that $\neg C_k(x) \,\&\, C_k(y)$, by R1. It appears then that the set of all characters predicated of y is a proper subset of that of all characters predicated of x.

$$\{A_1, A_2, \ldots, A_m, B_1, \ldots, B_n\} \supset \{B_1, \ldots, B_n\}.$$

The significance of R1 is that it ensures that the relation be that of proper set inclusion. This can be expressed in another way. Following Carnap,[19] let us define a 'state-description' of x as a conjunction

18 Rabbi Yeshuah Halevi ben Joseph, *Halikhot Olam* IV 2. These, as well as some further aspects of Talmudic logic are discussed in Heinrich Guggenheimer, 'Logical Problems in Jewish Tradition,' *Confrontations with Judaism*, ed. Philip Longworth (London, 1966), pp.171–96.

19 Rudolf Carnap, *Logical Foundations of Probability* (University of Chicago Press, 1962), p.70

which contains, for every attribute C_j predicable of x, either C_j or $\neg C_j$, according as the one or the other is true. Then a state-description of x and y is given by

$$x:\ A_1 \ \&\ A_2 \ \&\ \ldots \ \&\ A_m \ \&\ B_1 \ \&\ \ldots \ \&\ B_n.$$
$$y:\ \neg A_1 \ \&\ \neg A_2 \ \&\ \ldots \ \&\ \neg A_m \ \&\ B_1 \ \&\ \ldots \ \&\ B_n.$$

R1 means that there is nothing predicated of y that might be essential to the co-existence of all the B_i which cannot be predicated of x.
R2 means that there is no case where some A_i cannot co-exist with some B_j.

We can therefore safely posit an analogy between x and y that attributes to x all the positive characters of y, and that is precisely what the *kal va-ḥomer* does.

It is clear that the *kal va-ḥomer* does not give a certain inference. At best, the extension of the analogy will not lead to an inconsistency. Since the biblical text is a finite axiom system, it is possible, for a given *kal va-ḥomer*, to make sure that neither restriction R1 nor R2 is violated, and so avoid the danger of inconsistency. However, it is the very uncertainty that makes the method valuable. For the indeterminacy is due to the fact that the *kal va-ḥomer* enables us to say more than is deductively implied in the scriptural text. Moreover, this method of inference is not necessarily limited to scriptural interpretation alone. Furthermore, the Talmud recognizes the provisional use of a *kal va-ḥomer* inference, subject to later rejection if an invalidating instance is discovered.[20]

An example of this type of reasoning occurs in the following discussion about the binding rite of marriage, where the object is to establish that a marriage may be contracted by the groom giving his bride a coin or other object of value when it is the intent of both parties to make a marriage. It is known that cohabitation with intent to marry effects a legal marriage bond.[21] It is also explicitly provided in Scripture that a bondwoman can be acquired with coin and thereby a potential status of marriage is effected which can

20 *Sotah* 29b
21 Deuteronomy 24:1

become a regular marriage by a mere declaration.[22] Furthermore, the law of levirate marriage prescribes that the sister-in-law becomes the levir's wife by cohabitation alone, and not by any other rite.[23] On the other hand, the law makes it mandatory to release a bondwoman if she or others repay part of the purchase price before the potential marriage becomes actual,[24] while a normal marriage cannot be dissolved in this way.[25]

It is an inference: a Hebrew bond woman [y] cannot be contracted for by co-habitation [$\neg A(y)$] but can be contracted for by transfer of a coin [$B(y)$], this one [i.e., a wife] which can be contracted for by co-habitation [$A(x)$] should be capable of being contracted for by transfer of a coin.

$$[A(x) \mathbin{\&} \neg A(y) \rightarrow y < x.$$
$$y < x \mathbin{\&} B(y) \rightarrow B(x)]$$

It can be disputed [from the law of levirate marriage] that the sister-in-law [z] is contracted for by co-habitation but not by transfer of coin. [$A(z) \mathbin{\&} \neg B(z)$] [restriction 2] ... The premise of the inference can be disputed. Where do you infer from? From the Hebrew bondwoman? The Hebrew bondwoman can be released by the transfer of coin [$C(y)$], how then can you conclude regarding this one [a wife] who is not released by transfer of coin. [$\neg C(x)$] [restriction 1].[26]

The theory and ramifications of these kinds of inductive reasoning are highly developed in the rabbinic writings. In the Tanna-itic literature alone (i.e. up to and including the second century) approximately 900 instances of *kal va-ḥomer* have been counted,[27] and in the Talmud the number must run into thousands. For our purposes, it suffices to be aware of the central role of inductive processes in Talmudic logic.

22 Exodus 21:7–10
23 Deuteronomy 25:5
24 Exodus 21:8
25 Deuteronomy 24:1
26 *Kiddushin* 4b
27 Adolf Schwarz, *Der Hermeneutische Syllogismus* (Vienna, 1901) p.113

1.5 UNDECIDABLE PROPOSITIONS

One more feature of Talmudic logic needs to be mentioned, although it would take us too far afield to explain it adequately. That is the concept of *Teku*, literally 'let it stand.' Sometimes propositions are propounded whose truth or falsehood cannot be decided by use of any of the recognized modes of reasoning. When it appears that the existence of such a determination would lead to an inconsistency, the matter must be assumed to be undecidable and it is labelled a *teku*.[28]

Perhaps the simplest example of such an undecidable sentence in the Talmud is the following query raised by Rava. 'A nut is in a vessel and the vessel is floating on the water: do we follow the nut and [consider] it is at rest, or do we follow the vessel and that is not at rest for it is in motion?'[29] It is not too difficult to see why this is shelved with the remark '*Teku*.'

In a logical system where conclusions need not be demonstratively certain and where meaningful propositions can be undecidable, it is understandable that statistical inference can be admitted as legitimate. We shall see that in response to the need for decision-making, rabbinic logic did indeed develop the rudiments of probability theory.

28 See Rabbi Shlomo Luria, *Yam Shel Shlomo: Bava Kama* II 5
29 *Shabbat* 5b

2
Random mechanisms

2.0 CHANCE AND DIVINATION

The use of random mechanisms was widespread in pagan antiquity.[1] Games of chance were not only for entertainment: to pierce the veil of the future, divination by *astragali,*[2] freely falling arrows, and lots was frequent. In the administration of justice, too, the roll of the dice or some similar circumstance was often the deciding factor.

From the mathematical point of view, a 'fair' game is defined as one in which the various possible outcomes are equiprobable. To the extent that different random mechanisms are designed to assure 'fair' results, unaffected by skill or other considerations, they reveal at least an intuitive awareness of the concept of equal probabilities.

The Bible is vigorously opposed to the heathen practices of soothsaying and divination and includes among these abominations various types of recourse to chance. Thus Ezekiel tells us: 'For the king of Babylon stands at the parting of the way ... to use

1 An interesting collection of materials is found in F.N. David, *Games, Gods and Gambling* (New York, 1962), chaps. 1–4.

2 *Astragali* – knuckle bones taken from the hind-feet of sheep and used, like dice, for games of chance

divination; he shakes the arrows ... Into his right hand comes the lot for Jerusalem.'[3] And when wicked Haman sought to destroy the Jews, he chose the opportune time by lot. 'In the twelfth year of King Ahasuerus, they cast Pur, that is the lot, before Haman day after day and month after month till the twelfth month.'[4] Even when the procedure is not related to idolatrous worship the ancient rabbis make the condemnation of sortilege very explicit when they say:[5] 'Whence do we derive that one must not seek the guidance of lots? For it is written:[6] "You shall be blameless before the Lord your God [For these nations ... give heed to soothsayers and to diviners, but as for you, the Lord your God has not allowed you so to do]." '

Games of chance too were viewed with opprobrium, and gamblers were regarded as thieves. However, this is not primarily because they use gaming methods – rather it is because the winner takes his winnings without paying fair compensation. For this reason the Mishnah rules that the testimony of dice-players is unacceptable in a court of law.[7]

2.1 RECOURSE TO CHANCE IN THE BIBLE

On the other hand, the Bible and the Talmud record many instances where random mechanisms were used with approval, including some where reference to chance is mandatory. Besides the rite on the Day of Atonement, which will be discussed separately, these may be classified into two categories conveniently characterized by two passages in the Book of Proverbs.

A *'The lot is cast into the lap, but the decision is wholly from the Lord'*[8]

The concept of a completely chance occurrence implies indepen-

3 Ezekiel 21:26–27
4 Esther 3:7
5 Sifri quoted by Tosafot to Shabbat 156b *s.v.* כלדאי. However, in our printed text of the Sifri this passage does not appear.
6 Deuteronomy 8:13–14
7 *M. Sanhedrin* III 3
8 Proverbs 16:33

dence even from God and this is in contrast to the Jewish view of total Providence. Therefore, one must not dismiss the possibility that, under suitable conditions, the will of God may determine the outcome of a random choice, as a means of conveying a message. However, the rabbis are careful to point out that the lot reveals knowledge from God only when it is carried out by specific command and in connection with other approved procedures. This explains the use of lots by Joshua to discover that Achan had taken from the spoil of Jericho,[9] and by Saul to ascertain that Jonathan had broken the fast.[10] Furthermore, both Achan and Jonathan confessed their guilt, which was established by additional evidence also.

The uneasiness which the Talmudic sages felt about the reliability of lots as expressing the Divine will is poignantly expressed in the argument ascribed to Achan confronting Joshua: 'Joshua, do you want to charge me by a lot? You and Eleazar the priest are the two greatest men of the generation, yet were I to cast lots upon you, it must needs fall on one of you!'[11]

Thus it is clear that in the absence of evidence of heavenly sanction the outcome of the lot was considered to be pure chance.

B *'The lot puts an end to disputes and decides between powerful contenders.'*[12]

There are many cases where there are no criteria to decide justly between litigants whose rights are equal. In all such instances the use of random mechanisms is appropriate.

In the Bible, lots were used to insure (a) a fair division of property or privileges, as well as (b) duties and obligations. The following instances are noteworthy:

a /

1 The allocation of the land to the tribes of Israel and its apportionment among individual clans and families. This was commanded to Moses: 'The land shall be divided by lot';[13] and was carried

9 Joshua 7:16–18
10 Samuel 14:40–42
11 *Sanhedrin* 43b
12 Proverbs 18:18
13 Numbers 26:55

out by Joshua: 'These are the inheritances which the people of Israel received in the land of Canaan, which Eleazar the priest and Joshua, the son of Nun ... distributed to them. Their inheritance was by lot.'[14]

2 The assignment of towns to dwell in for the priests and Levites, who got no portion in the land: 'These are their dwelling places ... for theirs was the lot.'[15]

3 The order in which the priestly families were to officiate in the Temple: 'They organized them by lot';[16] and the appointments of the Levites to their various offices: 'These also ... cast lots:[17] They cast lots by fathers' houses, small and great alike, for their gates.'[18]

b / Conscripts for military and other duties were chosen by lots.

1 In the punitive action against Gibeah of Benjamin: 'We will go up against it by lot and we will take ten men of a hundred ...'[19]

2 To settle Jerusalem, when it was being rebuilt after the return from Babylonian exile: 'The rest of the people cast lots to bring one out of ten to live in Jerusalem the holy city, while nine-tenths remained in the other towns.'[20]

3 To bring wood-offering to the Temple: 'We have likewise cast lots, the priests, the Levites and the people, for the wood offering, to bring it into the House of our God.'[21]

The most commonly used term for lots in the Bible is גורל (*Goral*). By a natural development, it came to mean 'share' and 'fate.' While it occurs in the Bible several times in these latter senses,[22] in the Dead Sea Scrolls this usage is almost exclusive. However, there too, we find it used in connection with the apportionment of various labours by drawing lots. Thus we read in the

14 Joshua 14:1–2
15 I Chronicles 6:39 ff.
16 Ibid. 24:5
17 Ibid. 24:31
18 Ibid. 26:13
19 Judges 20:9–10
20 Nehemiah 11:1
21 Ibid. 10:35
22 E.g., Isaiah 17:14; Jeremiah 13:25; Psalms 16:5; Daniel 12:13; etc.

so-called *Manual of Discipline*: 'All heads of families within the community who are chosen by lot for communal service to go forth and come in before the congregation.'[23]

2.2 CHANCE MECHANISMS IN THE TALMUD

As if to emphasize their importance, the rabbis point out that four terms occur in the Bible for lots. 'The *goral* גורל is called by four names: *ḥelesh* חלש, *pur* פור, *goral* גורל, *ḥevel* חבל.'[24] In fact they explain additional instances in the biblical narrative where, although the Bible does not state it explicitly, lots were used. Furthermore, the Talmud tells of other instances where random mechanisms were used regularly, and it describes the actual procedures, so that the underlying mathematical insights can be ascertained.

We shall see that the rabbis were deeply aware of the need to neutralize such factors as might introduce a bias into a random procedure. To this end they designated various restrictions intended to prevent tampering with the lots. It was apparently recognized quite early that a fair procedure is one which, if repeated many times, gives each possible outcome with approximately the same relative frequency. This is a generalization from the special case involving just two choices, and we find it stated explicitly in the fifteenth century as a criterion for determining the effects of pure chance.

When there are only two possibilities the requirement is that, in a sufficiently long sequence of events, neither the one nor the other outcome should appear in a significant majority. If most of the events produce the same result, it cannot be ascribed to chance, but must be attributed to some causative agent or law. Now this concept already occurs in the Bible.[25] Job's friends maintain that the world is ruled by justice and the wicked are punished according to their deserts. To support this contention, they cite

23 *Manual of Discipline*, I 16: T.H. Gaster, *The Dead Sea Scriptures in English Translation* (New York, 1964), p. 328

24 *Pesikta de R. Kahana, Zakhor*, ed. Mandelbaum, vol. I, 36

25 Cf. Aristotle *Physica* II 8, 199b24, see below p.158

examples of the retribution meted out to sinners. Job demolishes this theory by a telling argument:

How often is it that the lamp of the wicked is put out?
That their calamity comes upon them? ...
That they are like straw before the wind and like chaff
that the storm carries away?[26]

Job does not question the veracity of his friends' reports about certain evil men who got their due, but he makes the point that since this is an infrequent occurrence it belies the existence of a moral order. If justice were the determinant of men's fortunes, then at least the majority of the wicked would be punished. Since that is not so, the opposite conclusion is warranted: 'The world is given over to the wicked.'[27]

Besides the relative frequencies of the various outcomes, the order in which they appear is also significant. We shall note that the regular recurrence of a particular pattern in the sequence of outcomes was indicative, to the rabbis, that design rather than chance was at work.

2.3 LOTS FOR THE SCAPEGOAT (*Goral*)

The Bible prescribes that on the annual Day of Atonement the High Priest shall offer two he-goats.

And he shall take from the congregation of the people of Israel two male goats for a sin offering ... and set them before the Lord ... and Aaron shall cast lots upon the two goats, one lot 'for the Lord' and the other lot 'for Azazel.' And Aaron shall present the goat on which fell the lot 'for the Lord,' and offer it as a sin offering; but the goat on which fell the lot 'for Azazel' shall be presented alive before the Lord to make atonement over it, that it may be sent away into the wilderness to Azazel.[28]

26 Job 21:17–18
27 Ibid. 9:24
28 Leviticus 6:3–10

From the detailed Talmudic description we quote: 'There was an urn there, and in it two lots. They were of box-wood until Ben Gamla made them of gold ... The lots may be of any material ... but they must be alike; They must not be one of gold and the other of silver, one big and the other small.'[29] The High Priest inserted both hands into the urn, bringing up one lot in each hand, and placed them upon the two animals, the one in his right hand upon the animal at his right and similarly for the left. Since it was regarded as an auspicious omen if the lot 'for the Lord' came up in the right hand, precautions were taken to prevent tampering with the lots. 'He [the High Priest] shook the urn and brought up the two lots ... Why did he shake the urn? In order to prevent choosing one intentionally ... The capacity of the urn ... was just great enough to contain no more than his two hands in order that he should not take one intentionally.'[30]

One of the sages suggested that the High Priest and his deputy each put in their right hand and bring up one lot only.[31]

It is described as a miracle due to the special merit of the high priest Shimon the Righteous that 'during the forty years that [he] served, the lot [''for the Lord''] came up in the right hand; thereafter, sometimes it came up in the right, sometimes in the left.'[32]

Later Bible commentators found in the rite of the lots for the scapegoat cause to discuss the meaning and role of chance. Thus Rabbi Isaac ben Mosheh Aramah writing in the fifteenth century explains:[33]

> That which is by chance is equal for everyone of the sides. While it is true that when Israel did God's will or when they had a worthy minister such as Shimon the Righteous, a miracle occurred with these lots in that the lot 'for the Lord' continually came up in the right ... this was by special providence and by the mouth of God as it is written 'The lot is cast into

29 *Yoma* 37a
30 Ibid. 39a
31 Ibid.
32 Ibid.
33 Rabbi Isaac ben Mosheh Aramah, *Akedat Yitzḥak*, chap. 63 (vol. III, 60b)

the lap, but the decision is wholly from the Lord.'[34] However later they were like ordinary lots that are due to chance without any greater tendency to one side rather than the other. Therefore, even if the lot 'for the Lord' were to come up in the right it would not be a 'sign,' for matters of this kind are not established unless they are found to be so many times ...

He is troubled by a seeming difficulty presented by the biblical record in which it is reported that when the ship in which Jonah fled came up upon a great tempest, the mariners said, 'Come, let us cast lots, that we may know on whose account this evil has come upon us.'[35]

Aramah comments:

Apparently this was foolish counsel. For it is impossible for it to be otherwise than that the lot should fall on one of them whether he be innocent or guilty, as the rabbis explained concerning Achan ...[36] However the meaning of their statement 'let us cast lots' is to cast lots many times. Therefore the plural – *goralot* – is used rather than a [single] *goral* cast on many people as in the case where it is said: 'the land shall be divided by lot [*goral*].[37] They did so and cast lots many times and every time the lot fell on Jonah and consequently the matter was verified for them. It follows then that the casting of a lot indicates primarily a reference to chance.

2.4 DIVISION OF THE HOLY LAND

The Bible relates: 'Joshua cast lots for them in Shiloh before the Lord; and there Joshua apportioned the land to the people of Israel, to each his portion.'[38]

In the Jerusalem Talmud we are told:'With two urns was the land of Israel divided – one containing lots [with the names of the district] and the other the names of the tribes. Two young priests stood by – what this one drew and that one drew was assigned.'[39]

34 Proverbs 16:33
35 Jonah 1:7
36 *Sanhedrin* 43b. See above, p.23, where the relevant passage is quoted
37 Numbers 26:55
38 Joshua 18:10
39 J. *Yoma* IV 1

Here we have a more elaborate coincidence method, which effectively precludes cheating. This, together with the fact that the portions assigned were of equal value, ought to have been enough to satisfy everybody. Nonetheless, the Talmud adds that the final sanction of the division of the land was by prophetic inspiration.[40]

Based on this precedent, Rabbi Yosé ruled that brothers who divided their patrimony into equal shares may, by unanimous consent, allocate, the portions by lot.[41] Although, in such a case, there would, of course, be no prophetic sanction, it was held that mutual agreement suffices to validate the outcome.[42]

2.5 REDEMPTION OF THE FIRST BORN

The Bible tells:

> The Lord said to Moses, 'Take the Levites instead of all the first-born among the people of Israel ... and the Levites shall be mine: I am the Lord. And for the redemption of the two hundred and seventy-three of the first born of the people of Israel, over and above the number of the male Levites, you shall take five shekels apiece ... and give the money by which the excess number of them is redeemed to Aaron and his sons.' So Moses took the redemption money from those who were over and above those redeemed by the Levites.[43]

The Talmud comments:

> Moses said, 'How shall I do it to the Israelites? Should I say to one: "Give me your redemption money and be discharged," he will say to me, "I have already been redeemed by exchange with a Levite."'[44]

The solution is, of course, by lots.

> He brought 22,000 ballots and wrote on them 'Levite,' and 273 on which

40 *Bava Batra* 122a
41 *T. Bava Batra* III 7
42 *Bava Batra* 106b
43 Numbers 3:44–49
44 *Sanhedrin* 17a

he wrote 'five shekels.' He mixed them up thoroughly, put them in an urn, and said to them, 'Take your ballots.' Whoever drew a ballot 'Levite' – to him he said, 'You have been redeemed already by exchange with a Levite.' He who drew a ballot [marked] 'five shekels' - to him he said, 'Give your redemption money and be discharged.'[45]

However, the Jerusalem Talmud records that there were misgivings on the fairness of the lot reminiscent of Achan's claims, so that proof of divine approval was needed to make it binding.

Rabbi Judah and Rabbi Nehemiah – one asked the other: [One could object] 'Had you written a "Levite" for me, that would settle it [i.e. since there were less ballots marked "Levite" than people drawing, I did not have a chance].'

He [Moses] did thus, then. He took 22,000 ballots and wrote on them 'Levite,' and 273 more and also wrote on them 'Levite,' and another 273 and wrote on them 'five shekels' and put them in an urn etc. One asked the other: Suppose that everybody drew 'Levite?'

He answered: It was a miracle that they came up at regular intervals [i.e. all 273 ballots marked 'five shekels' were drawn and they came up at regular intervals].

Said Rabbi Samuel: According to the latter teacher, there was a miracle, but according to the former [who taught that the total number of ballots equalled the number of people] was there no miracle? [What divine sanction was there then for this procedure?] [Rabbi Abbahu] said to him: In any case, there was a miracle since they alternated at regular intervals.[46]

While one cannot dismiss the possibility that if 22,273 ballots in all were to be drawn from a total of 22,546, at least one, say, of the 273 marked 'five shekels' might remain in the urn, yet neither is it unlikely for all of them to be drawn, since the density in such a large sample would be only slightly greater than in the original mixture. In fact, both possibilities are well within the expected range of variation. Therefore, the elimination of one of them (i.e.

45 Ibid.
46 *J. Sanhedrin* I 4

no 'five shekels' left in the urn), can hardly be attributed to a miracle. Thus, the suggestion of a miracle could not be made without adding the additional feature of alternation at regular intervals. The probability of such an occurrence is exceedingly small and it must therefore be attributed to an extraneous cause. That is the point made in the remarks of Rabbi Samuel and Rabbi Abbahu, that a miracle must be postulated only for the regular alternation.[47] It is reminiscent of Aristotle's argument that 'to throw ten thousand *coan* throws [sixes] with the dice would be impossible, but to throw one or two is comparatively easy.'[48]

2.6 SHARING THE FLESH OF THE SACRIFICIAL ANIMALS (*Ḥelesh*)

According to the Mishnah,[49] lots were used regularly to divide the meat of the sacrifices among the priests after it was cut up into portions.

Maimonides, in his commentary, explains the procedure used, but his sources are unknown to me. He writes:

> They would bring different-looking objects numbering as many as the people and each one would choose one of these objects as his own. Then they would hand over these [markers] to another person who was unaware of how they had assigned them amongst themselves, and he would place each of these things [i.e. markers] on a portion of meat.

47 A similar situation is described with respect to the appointment of the seventy elders (Numbers 11:16). The Talmud suggests that Moses actually selected six from each of the twelve tribes, making seventy-two in all, and then eliminated two by lots. The entire analysis is paired with that on the redemption of the first-born, both in the Babylonian and Jerusalem Talmuds. Because of the exact parallelism, the quotations in the text are limited to the discussion on the first-born.

48 *De Caelo* II 12,292a30

49 *M. Shabbat* XXIII 2. The Mishnah uses the term *ḥelesh* for this procedure. However, in Sifri on Numbers 18:7, the word used is *payis*, a term which is discussed in the following section.

2.7 ASSIGNMENT OF PRIESTLY DUTIES: THE LAW OF LARGE NUMBERS

A different and much more interesting type of random mechanism was used in assigning the daily duties of the priests in making the Temple offerings. In fact, this procedure was called by a special name – *payis*, which also means 'to conciliate or pacify.'[50] At one time, certain priestly duties were assigned on a 'first come, first serve' basis. This led the priests to race each other up the ramp to the altar, and once one of them broke his leg as a result. Then the *payis* was instituted for all functions.[51] Thus, thirteen priests were required for the daily offering, and the *payis* was designed so that the priest who was assigned the daily offering found himself leading twelve others in a specified order.

Rashi, in his commentary to the Talmud, describes the procedure concisely as follows:

> They all stand in a circle. The officer removes the mitre from one of them and the counting will begin with that one. Each one puts forward a finger to be counted and the officer says, 'He wins upon whom the counting will end.' He announces a number ... much greater than the number of priests present and begins to count from him whose mitre was removed and continues around and around until the preassigned number is reached.[52]

The counting was done on the fingers rather than on the priests themselves because it was forbidden to count Israelites. However, this created a problem of another kind. Elderly priests would find it difficult to keep one finger alone extended for a long time, and so permission was granted to them to extend two fingers.

Obviously, if those people who extended two fingers were to be counted twice, the outcomes would not be equiprobable. Now the Talmud explains: 'The sick and the venerable put forward two [fingers] but they are counted as one only.'

50 Marcus Jastrow, *Dictionary* ... , 2 vols., New York, 1950, 1162 s.v. פייס and פייסא

51 *Yoma* 22a

52 Ibid.

On the other hand, to avoid cheating, the following rule was laid down: 'One does not put forth either the third finger or the thumb, because of tricksters. If one put forth the third finger it is counted. However, the thumb is not counted; moreover, the officer strikes him ...'[53] This was because, towards the end of the count when the outcome could already be gauged, a trickster might put forward both thumb and finger spreading them so far apart as to mislead the officer into counting them as two fingers belonging to two different people.[54]

The mathematical basis of the method is not explained. We have here first a random choice of the person from whom the counting is to begin and on the first count around, each priest gets a number k, between 1 and m, where m is the total number of priests. Since this choice does not determine anything as yet, there is no reason to suspect that it is other than by pure chance, and so every priest has an equal chance to get any number k. Then we have, selected at random, a large number N.

Each complete circuit of the priests adds m to the count, so that, if b is the largest number of complete circuits that can be made, $(N - bm)$ gives the number of the winning priest. The remainder $(N - bm)$ is called the residue of N modulo m. N is a random variable, that is, any positive integer, between very wide limits, has an equal chance of being picked. Therefore, its residue modulo m is equally distributed over the numbers $1, 2, \ldots, m$. The probability

53 Ibid. 23a

54 This discussion is based on the reading of our standard printed text, as explained by Rashi, and supported by the reading of the *Tosefta Yoma* I 9. However, Maimonides' description of the *payis* in his Code (Daily Offerings IV 4) differs in crucial points and is based on a variant reading which does not appear in our Talmudic texts. This is clear from his *Commentary to the Mishnah, Yoma* II 1 and his responsum on the subject (ed. Blau, no.126). See also the remarks of J. Kapaḥ in his edition of Maimonides' *Commentary of the Mishnah* (Jerusalem, 1964) vol. 2, 156 nn. 5 and 6; also *Otzar Ha-Geonim*, ed. Lewin (Jerusalem, 1934), vol. 6 (*Yoma*), 9 and 51 where similar and other variants are found. Hasofer ('Random Mechanisms in Talmudic Literature,' *Biometrika*, 54 [1966] 318 ff.), who first drew attention to the mathematical significance of the *payis*, was apparently unaware of the textual differences and quotes Maimonides inaccurately.

that a given priest will win, is compounded of the probability that he will be chosen to be number k and the probability that the number N will be chosen which has a residue k modulo m. Since these probabilities are equal for all k between 1 and m, every priest has an equal chance.

It remains to point out that with respect to the burning of incense, a highly valued duty, it was announced: 'Those new to the incense, come and draw.'[55]

Usually, all the priests partook in the *payis,* but in this case, sampling *without* replacement was used to make sure that nobody got a second chance before everybody had had a first one. It was, then, clearly understood that as a random method, the *payis* must be expected occasionally to give one the edge over another.

The priests served in shifts in a cycle of twenty-four weeks, so that the same priests would minister every day for one week and then be discharged for twenty-three weeks.[56] Albeit, since the *payis* was repeated four times daily, and the priests were very jealous of their privileges, any bias in the method of assignment would come to light after several years at most. The *payis*, however, seems to have been accepted gracefully, in fact the term acquired the meaning of pacification. This would indicate that, in spite of the inevitable vagaries of chance which might give the odd person a streak of luck, it was realized that in the long run the distribution was fair. Thus we have an intuition of a special case of Jacob Bernoulli's celebrated theorem known as the 'Law of Large Numbers,'[57] namely, that as the number of trials increases, equality of distribution is approached, i.e., the discrepancy between the number of wins by each of the priests from his fair share, averaged over the number of trials, will decrease towards zero. In other words, a 'fair' method really produces a distribution close to equal

55 *Yoma* 26a

56 I Chronicles 24:7–19, M. *Taanit* IV 1

57 Let p be the probability of an event, m the number of occurrences of the event in n trials, and e an arbitrarily small positive number. If P is the probability that the inequality $|(m/n) - p| < e$ is satisfied, then $\lim_{n\to\infty} P = 1$.

The theorem was given by Bernoulli in his *Ars Conjectandi* published posthumously (Basle, 1713. Photo-offset; Brussels, 1968), p.236.

only when used many times. As already mentioned above (§2.3) this insight was made explicit in the fifteenth century by Rabbi Isaac Aramah[58] who thus became the earliest known author to whom Bernoulli's theorem can be credited. For Aramah makes it clear that where the outcome is due to chance alone, in many trials all possibilities should be realised in equal proportion.

58 Although Aramah's formulation is more explicit, it is closely related to an earlier statement by Maimonides, which will be examined later (p.75) because it occurs in a different context.

3

Follow the majority: a frequency interpretation

3.0 ACCEPTANCE RULES

One of the simplest problems in elementary textbooks of mathematical probability is of the following kind. Given an urn containing m white balls and n black ones, what is the probability that a ball drawn from the urn will turn out to be white? Such problems may arise among gamblers who wish to compute what stakes it is reasonable to put on the next draw, but hardly in any other context.

It is, however, only a minor modification of this problem to consider the following. An urn contains $(m + n)$ objects, all of which look alike but concerning which it is known that m are sound while the remaining n have some internal defect. If I draw one object from the urn, what is the probability that it is sound? Such situations occur frequently in most kinds of manufacturing plants and the question is raised in connection with the practical decision problem of whether a given product is to be accepted or not. An 'acceptance rule' is one in accordance with which we decide to approve a product if the probability that it is sound is greater than some fixed number. In our hypothetical example, the ratio of sound objects to the total number in the collection is $m/(m + n)$, and one defines the probability $p(s)$ that an object selected from the urn is sound as

$$p(s) = m/(m + n).$$

It follows that, if $m = 0$, all the articles are defective and $p(s)$ takes its smallest possible value, namely, zero. On the other hand, if $n = 0$, all the articles are sound and $p(s)$ takes its maximum possible value, namely, one. Depending upon the relative magnitudes of m and n, $p(s)$ can take various values between zero and one. Therefore, an *acceptance rule* will be of the form: accept if $p(s) > k$ (read $p(s)$ is greater than k) where k is some number between zero and one. Of course, one can speak also of a corresponding 'rejection rule' which specifies when we reject a given article, and in many applications it is more convenient to use this form of rule. In most situations, one would not want to accept a product if the chances are even or worse that the product is sound. If we are satisfied with something slightly better than even chances, namely $p(s) > 1/2$ or $k = 1/2$, we can designate our rule as a *straight majority acceptance rule*. Otherwise we may require k to be closer to 1. How close to 1 we shall want it to be will depend mainly upon what we anticipate to be the results of making a mistake. Obviously, if the article under consideration is such that our lives may depend upon its being free from defect, we should be prepared to undertake the extra cost involved in ascertaining beyond doubt whether it is sound.

In general, there are two possible kinds of error to be avoided. On the one hand, one might approve a defective, thus incurring a possible loss when it fails to serve as expected, and on the other hand, one might reject a sound product, thus incurring a loss insofar as the cost of production is concerned.

The same problems can be viewed in a somewhat different light. Rather than as a practical question of accepting or rejecting a product, it can be seen as a matter of making inferences about the state of nature. In our example, there are two possibilities – either the article is defective or it is sound. These correspond to two hypotheses, say H_0 and H_1. We now formulate a 'rule of inference' that says: 'If $p(s) > k$, conclude H_1; otherwise, H_0.' This *rule of inference* may also be called an *acceptance rule* since the two are defined equivalently.

To characterize the effectiveness of these rules, 'significance level' is defined as a measure of the risk one is prepared to take that the rule of inference will lead to the rejection of an hypothesis that is true.[1]

3.1 FOLLOW THE MAJORITY

In the Talmud, acceptance rules are defined in situations much like those involved in the decision problems described above. The most common is a straight majority rule. This is derived from a biblical provision with respect to the judiciary.

According to the Bible[2] and the Talmud,[3] both judicial and legislative authority are vested in the supreme court, called the Great Sanhedrin, consisting of seventy-one members. Lower courts, with jurisdiction in capital cases have twenty-three members, while panels of three judges are empowered to hear civil and monetary cases. Decisions are reached by a simple majority, except that for conviction in capital cases a plurality of at least two is required. The rule 'Follow the Majority,' which in the first instance applies to judicial procedure, was extended by the rabbis, who interpreted it as the fundamental principle of statistical inference: 'In the entire Law we adopt the rule that a majority [or the larger portion] is equivalent to the whole.'[4]

3.2 COUNTED AND UNCOUNTED MAJORITIES

A typical question involves the legal status of objects and persons when their provenance is unknown. For example, are they permitted or forbidden, ritually clean or unclean, etc? Thus, only meat which has been slaughtered in the prescribed manner is *kasher* – permitted for food. If it is known that most of the meat available in a town is *kasher*, there being, say, nine shops selling *kasher* meat

1 D.J. Finney, *Statistics for Mathematicians* (Edinburgh, 1968), p.47

2 Numbers 11:16

3 *Sanhedrin* I, 1–6

4 *Horayot* 3b. The biblical source, Job 21:17, indicated above in §2.2 is not referred to in this connection: authority for legal practice must be found in the Pentateuch rather than in the other books of Scripture (*Ḥagigah* 10b).

and only one that sells non-*kasher* meat, then it can be assumed when an unidentified piece of meat is found in the street that it came from the majority and is therefore permitted.

The same rule of following the majority applies in many different cases. However, the Talmud distinguishes between an enumerated majority and one that is just taken for granted a priori:

> Whence is derived the rule which the rabbis stated: 'Follow the Majority?' ... As for a majority that is enumerated, as in the case of the Nine shops or the Sanhedrin, we do not ask the question. Our question relates to cases where the majority is not explicit as in the case of the Boy and Girl [who, we assume, will grow up to be fertile and are therefore not to be considered eunuchs with respect to levirate marriage, although this cannot now be ascertained. The reason is that we follow the majority, and the majority of people are not sterile. Yet, this majority is not an explicit one since we cannot count all the people].[5]

Not all the sages agreed to follow a majority when it is not actually enumerated. Rabbi Meir and others 'are concerned for the minority'[6] except where the uncounted majority is overwhelming. Such is the presumption that most of 'the scribes of the courts know the law,'[7] for an ignorant scribe must be so rare that a legal document issued by the court may be assumed to have been properly executed.

3.3 TALMUDIC ACCEPTANCE RULES

'Follow the Majority' is justified by the assumption cited in the Talmud: 'That which is separated [from a collection] derives from the majority.'[8] In fact, we are told that 'by Biblical law, one in two is nullified,'[9] that is, if one forbidden object is mixed with two per-

5 *Ḥullin* 11a
6 *Gittin* 2b
7 Ibid.
8 *Berakhot* 28a, *Yoma* 84b, etc.
9 *Gittin* 54b

mitted ones, the prohibition is annulled. This is understood by most of the commentators in the sense of what would be called today an acceptance rule with $k = 2/3$. Rabbi Shlomo ben Adret explains that, in the case of meat, say, each piece can be drawn separately, and so 'When he eats the one, I say "This is not the forbidden one" and similarly for each piece. When he eats the last piece I say "The forbidden one was among those already consumed and this one is permitted" ... since by Biblical law one in two is nullified.'[10]

It is noteworthy that a likely cause for confusion is here mentioned and eliminated. After the first piece is drawn, is the probability still 2/3 that the second piece is *kasher*? Clearly, if the first piece were identified, that would not be so. For if that piece were found to be the prohibited one, the remaining two pieces certainly would then be *kasher*, while if proved to be *kasher* they would be half and half. However, since no additional information is forthcoming as a result of drawing the first piece, the probabilities do not change. When the last piece is taken, the probability is still 2/3 that it is permitted and by the given acceptance rule it is inferred that this piece is *kasher*, which implies that 'the forbidden one was among those already consumed.'

Frequently, the Talmud speaks merely of nullification in a majority[11] which corresponds to a straight majority acceptance rule; occasionally, though, the phrase used is 'one in two is nullified,'[12] which seems to imply a need for a 2:1 ratio. Since the legal principle involved is ultimately connected with the judicial procedure where, in the tribunal of seventy-one, a split of 36:35 can determine a verdict, some authorities do not consider the 2:1 ratio as a prerequisite; they maintain that its mention is motivated only by the desire to use the simplest possible numbers.[13] In a mixture of any number of individual objects, the forbidden minority being unrecognizable, even a simple majority of one suffices, by biblical

10 *Torat HaBayit HaArokh* IV 1 (68b)
11 E.g. *Ḥullin* 100a
12 *Gittin* 54b
13 Rabbi Shlomo ben Adret, *1255 Responsa* No. 272

law, according to this view, to nullify the prohibition.[14] However, it seems that Maimonides[15] and others regard the 2:1 ratio as a minimum requirement when the status of the entire mixture is at issue, namely, when all the elements will be drawn, albeit one at a time. One reason is that in a mixture the forbidden element is certainly present. Thus, following the majority for every drawing implies that eventually an error will be made and a forbidden object will be permitted, and to actually annul a prohibition only a decisive majority will do. However, if the question does not concern the complete mixture; for example, if some of it is not accessible or is no longer extant when the matter is decided, then even a simple majority may suffice, since a decision on those elements with which we are concerned may in fact never involve error at all.[16]

In practice, the Talmud itself distinguishes between prohibitions of differing gravity for which, at least by rabbinic law, considerably higher majorities are required for annulment. Essentially, we have here a concept of 'significance level,' which measures the risk one is prepared to take that 'Follow the Majority' will result in error. There are seven such levels[17] which can be adequately illustrated by a few examples.

Terumah is an offering to be set aside from the grain crop. It is sacred and may be eaten only by priests under exacting conditions of ritual purity, so that a mixture of ordinary grain and *terumah* is forbidden to all except priests, unless the ratio is 100:1. Other more severe prohibitions require ratios of 200:1,[18] 1000:1,[19] and, in the case of objects used in idolatrous cults, even only one in ten thousand prohibits the entire mixture.[20] With respect to certain kinds of poisons, such as snake venom, the Talmud rules that even the faintest possibility that a trace of venom is present renders a mixture forbidden, for 'danger is graver than prohibition.' Thus

14 Rabbi Bezalel Ashkenazi, *Shittah Mekubetzet* to *Betzah* 4a s.v. ר' יהושע
15 Moses Maimonides, *Mishneh Torah, Forbidden Foods* XV 4
16 See Rabbi Shlomo ben Adret, *Torat haBayit haArokh* IV 2(75a)
17 See Maimonides, *Mishneh Torah, Forbidden Foods* XV
18 *J. Orlah* II 1, see also *Ḥullin* 99b, 100a
19 *J. Terumot* X 8
20 *Zevaḥim* 74a, based on Deuteronomy 13:18

even if nine people were unharmed after drinking from a barrel which was unprotected from access by snakes, nonetheless one may not drink from it, for it is possible that there might yet be venom there.[21]

Rashi points out that following the majority always involves 'uncertainty.'[22] Yet, this uncertainty is reduced as the relative frequency of the majority elements is increased. Clearly, the amount of uncertainty that may be allowed depends upon the nature of the consequences that might follow from acting on a decision that is not in accordance with the facts. This, then, determines what is a suitable acceptance rule.

3.4 APPROXIMATE FREQUENCIES

A precise measure of relative frequency is usually possible only when dealing with an enumerated collection, such as in the examples cited of mixtures. In many situations, however, one must deal with unenumerated majorities. In one respect, for unenumerated classes greater confidence is justified in using an acceptance rule. Usually a decision does not inevitably involve error, since in such cases a ruling applies only to a few individuals, many being inaccessible. There still is, of course, some risk of error, but it is not a contradiction to suppose that no mistake will be made, since it is possible that for all those elements which come up for decision, the inference rule will give true results. On the other hand, there is a difficulty in prescribing an appropriate acceptance rule, for only rough and approximate levels of relative frequency can be ascertained in uncounted collections. This gives only a few kinds of majorities which are then ordered comparatively.

We have already seen that certain majorities are so overwhelming that they are taken as practical certainties (the example of the court-scribes, whose appointment is subject to responsible authorities). There is a class of such legal presumptions, where, because there are known factors at work to produce and maintain a given situation, the assumption of certainty is justified.

21 *Avodah Zarah* 30b
22 *Pesaḥim* 9a. s.v. ודאי וודאי

In other contexts we come across situations where exceptions do occur but are nonetheless rare enough to warrant near certainty. Thus, in a case of an aborted foetus that was cast into a pit 'the sages ruled it [the pit] clean [from the ritual defilement of the dead] because weasels and martens frequented it,'[23] although there is 'a doubt whether they had already dragged it [the foetus] away or not.' In a gloss of the Tosafot,[24] a picturesque term is used to describe the situation – 'a usual doubt,' meaning that the one alternative is usual rather than the other, for it is quite unlikely that animals will not have dragged away the remains. Another level is a 'significant' majority which is distinguished from just an ordinary one.[25]

A barely acceptable majority is not, or under most circumstances can it be, defined for uncounted collections. Thus a precise acceptance rule cannot be formulated.

3.5 STANDARD MAJORITIES

Many different examples occur in the Talmud of 'standard' majorities, and it is clear that in order to be recognized as such directly, they must be rather substantial.[26] Some examples follow:

1 'most women's pregnancies last nine months';[27]
2 'for most women who bear at nine months, pregnancy is already recognizable at one third term [that is three months]';[28]
3 'most children will turn out to be fertile';[29]
4 'most people who buy an ox intend it for ploughing [work]';[30]

23 *Pesaḥim* 9b
24 Ibid. s.v. ספק and 9a s.v. ואם where it is explained 'a usual doubt – close to certainty.' See also Tosafot to *Avodah Zarah* 41b s.v. ואין
25 *Tosafot Kiddushin* 80a s.v. סמוך
26 See W. Kneale, *Probability and Induction* (Oxford, 1949), p.194: '... we have a capacity for forming impressions of the relative size of groups or sequences whose members we have not counted.'
27 *Yevamot* 37a
28 Ibid.
29 *Yevamot* 119a
30 *Bava Kama* 46b

5 'most cattle are *kasher* [that is, do not have disqualifying diseases]';[31]
6 'most lions claw at their prey';[32]
7 'most cattle do not lactate before giving birth.'[33]

Majorities arrived at by computation will be discussed later.

Perhaps a lower limit can be ascertained for what would be considered an overwhelming majority from the following text. In a discussion concerning cheese made with rennet derived from the stomachs of cattle we read: 'Since there would be a minority of calves that are slaughtered for idolatry and then there are the other [adult] cattle [of whom none are slaughtered for idolatry and they are a majority as against all calves] it would be a minority of a minority and even Rabbi Meir is not concerned for a minority of a minority.'[34] It is clear that a 'minority of a minority' cannot be as great as one quarter. On the other hand, at least one of the commentators felt that, unless it is known to be a minority of a minority, an unenumerated minority cannot be assumed to be small enough.[35]

3.6 HALF AND HALF

Wherever there is no clear majority, the existence of a doubt is regarded in the Talmud as 'half and half.'[36] Similarly, in a dispute between two litigants, where the respective probabilities of their claims are equal, the property at issue is divided between them.[37] It is noteworthy that here we have a very early example of the convention that assigns to the range of the probability function the interval (0,1).

3.7 RELEVANT REFERENCE CLASSES

To define an acceptance rule, it is always necessary to specify a

31 *Ḥullin* 11a
32 *Ḥullin* 53a
33 *Bekhorot* 20b
34 *Avodah Zarah* 34b
35 Rabbi Yom Tov ben Abraham, *Commentary to Niddah* 32a, s.v. קטן
36 *M. Makhshirin* II 3–11
37 *Bava Batra* 93a, see commentary of Rabbi Shlomo ben Meir s.v. ורביע

reference class. If, as was explained above, all the elements of this class are supposed to have one of two contradictory properties, say *A* and not-*A*, it is then possible to define the ratio of the number of *A*s to the total number in the given reference class (the relative frequency) and to formulate an acceptance rule by prescribing a minimum for this ratio. When the reference class is understood, it need not be mentioned explicitly. Thus, in the case cited above about the pit that was frequented by weasels, we might explain that the reference class consists of all visits by weasels to the pit and in a majority of these the animals eat or drag away any flesh they might find there. Unless it is clear, though, which reference class is meant, the definition of the acceptance rule is incomplete.[38] The rabbis raised this issue when ambiguity could arise about which reference class was relevant.

The determination of whether the majority is relevant to the case in question seems to be at the root of the following argument:

All that is stationary [fixed] is considered as half and half ... If nine shops sell ritually slaughtered meat and one sells meat that is not ritually slaughtered and he bought in one of them and does not know which one – it is prohibited because of the doubt; but if meat was found [in the street] one goes after the majority.[39]

Although a dogmatic motive is given for this distinction, in terms of a biblical ordinance, and the commentators differ as to its precise scope, the reasoning seems to be that when the question arises at the source, the chances are not really nine to one. For the other nine shops do not enter into the picture at all, since the piece of meat in question certainly does not come from any of them. Therefore there are only two possibilities and the chances of its being *kasher* or not must be considered even.

In other passages, the point is explicitly made that there are different classes which can contain the elements in question, and it may be that different probabilities can be attributed to the same

38 Compare Karl R. Popper, *The Logic of Scientific Discovery* (Harper, 1968), p.210

39 *Ketubot* 15a. See Rabbi Nissim ben Reuven to *Sanhedrin* 79a, who argues that in one view the distinction is of rabbinic origin only

element according as one or the other of the reference classes applies. Thus:

One sold grain to his fellow who planted it and it failed to grow ... in the case of flaxseed, the seller is not liable. Rabbi Yosé says: He must give him the value of the seed. They [Rabbi Yosé's colleagues] said to him: Many buy it for other purposes [for food and medicine].

... Both of them [Rabbi Yosé and his colleagues] follow the majority – one goes after the majority of people [who buy flaxseed for other purposes] and one goes after the majority of seed [which is mainly used for sowing].[40]

The event we are concerned with here is that of a man who purchased a certain quantity of seed. If the determinant feature of this event is the one man who is buying, the reference class is that of all men who buy flaxseed. A majority of this class buy seed for a variety of purposes which do not require the seeds to be fertile. Rabbi Yosé thinks that one ought to look at this event as a transfer of a quantity of seed and the appropriate reference class is that of all flaxseed that changes hands. Since a majority of seed bought is intended for planting, fertility is a crucial consideration.

There are cases in the Talmud where two different classes may be equally relevant and majorities in both classes are required for a decision.[41] Where one class is contained within the other, the smaller class only is to be taken into account. Thus: 'Barrels of wine were found in Be-Khefe. Rava permitted them [although most of the inhabitants were non-Jews and wine made by them is prohibited] ... It is different here, because the majority of wine-dealers were Jewish: this is so only for large barrels [such as wine dealers use].'[42]

Rather than the larger class of the total population, or the wine production, only the subclass of wine merchants is the suitable reference class.[43]

40 *Baba Batra* 93b

41 *Ketubot* 15a

42 *Bava Batra* 24a

43 For a temporal sequence of events, the same rule is applied in *Pesaḥim* 7a with a reference to monies found in Jerusalem where the streets were swept daily and so one does not consider the larger class of all visitors as potential losers, only the more recent ones.

3.8 A FREQUENCY INTERPRETATION

At the beginning of this chapter the probability $p(s)$ was defined of drawing a sound object from an urn containing m sound and n defective articles. On this basis an acceptance rule was defined. Crucial to these definitions is the reference class. Whether and in what context a reasonable definition of probability can be constructed without a reference class is a question that still exercises philosophers and logicians. However, when a finite reference class is available, it is universally agreed that the definition of probability given earlier, namely

$$p(s) = m/(m + n),$$

is acceptable and meaningful. Some have maintained, though, that only a frequency interpretation is at all admissible, and where no suitable reference class is available, probability statements can have no meaning. The frequency interpretation dates from at least 1842, when R.L. Ellis argued that 'It seems to me to be true *a priori* ... that one event is more likely to happen than another [means] that in the long run it will occur more frequently.'[44] The theory was developed by John Venn in his classic work *The Logic of Chance*. He declared: 'By assigning ... an expectation in reference to the individual, we mean nothing more than to make a statement about the average of his class.'[45]

It is clear from the Talmudic texts quoted that the rabbis conceived of their acceptance rules in terms of frequencies, and they were careful to consider the appropriateness of the requisite reference class. In terms of frequencies, they defined acceptance rules with varying significance levels. We shall see that they also developed what may be called an arithmetic of probabilities on the basis of relative frequencies.

44 R.L. Ellis, 'On the foundation of the theory of probabilities,' *Trans. Camb. Phil. Soc.* vol. 8 (1849), 1 ff

45 John Venn, *The Logic of Chance*, 3rd ed. (London, 1888), p.151

4
Addition and multiplication of probabilities

4.0 ARITHMETIC FOR ACCEPTANCE RULES

It has been shown that the Talmud formulates acceptance rules defined in terms of relative frequencies. Thus the basic notion is an acceptance rule. In the modern theory, one first defines the notion of probability as identical with relative frequency, and then acceptance rules are framed in terms of probabilities. Whereas addition and multiplication are more easily defined for probabilities, the same operations with respect to acceptance rules are somewhat complicated, since in the computations relative frequencies must be used while the results are given in terms of acceptance rules. The rabbis proceeding from the need for decision procedures had recourse to such arithmetical operations to define acceptance rules in cases where, although majorities cannot be directly ascertained, yet it may be possible to determine by computation whether there exists a relevant majority which warrants the application of an acceptance rule.

4.1 ADDITION

Let us consider first addition. In the case of known relative frequencies such as measured mixtures where precise proportions are given, addition is quite straightforward, as can be seen from the following example. '*Terumah*, and the *terumah* set aside from the doubtful tithe and *ḥallah* [first offering from dough] and first-fruits are each neutralized [in a mixture] of one to a hundred, and they are added one to another [that is, if varying quantities of each fell into ordinary grain, the amounts of the forbidden grains are added up and if that is 1/101 or less of the total mixture, it is all permitted; otherwise it is prohibited].'[1]

What is sought is the relative frequency of the set which consists of the sum of various individual sets – in the above example, the union set which consists of all elements that are either *terumah*, or *terumah* from doubtful tithe, or *ḥallah*, or first-fruits. This is equivalent to adding up the individual relative frequencies of each of these sets, and that can easily be obtained by adding together the quantities of each of these types of forbidden fruits and then determining the ratio of that sum to the entire mixture.

The Mishnah is careful to point out right at the beginning that the acceptance rule with respect to admixtures of all of these prohibited items is the same, namely, each is neutralized in a ratio of one to a hundred. The problem of stating an acceptance rule for a mixture containing more than one kind of such prohibited things reduces to adding up the individual probabilities, as we have seen.

The matter is much more complicated if the prohibitions are of different types for which different acceptance rules apply. This raises a series of legal questions beyond the mathematical ones, but it is important to note that the purely probabilistic considerations were kept clearly distinct from the legal ones. In our example, in addition to *terumah* for which a ratio of 1:100 is required, one other prohibition appears, namely, *orlah*. If a tree during its first three years bears any fruit, it is known as *orlah* and is prohibited not only for food but for any use whatsoever.[2] Unlike *terumah* which is permitted to priests and prohibited to all others, *orlah* is forbidden to all alike. The acceptance rule is: a mixture containing no more than one part of *orlah* to 200 parts of ordinary fruit is permitted.

1 *M. Orlah* II 1
2 Leviticus 19:23

The Mishnah discusses the status of a mixture containing both *terumah* and *orlah*: '*Terumah* can neutralize *orlah* ... How is it? One *seah* [a dry measure] of *terumah* fell into [ordinary fruit to make up] a hundred, and then three *kab* [= 1/2 *seah*] of *orlah* fell [into the mixture] ... In this case the *terumah* neutralizes the *orlah*.'[3]

The Gemara[4] explains that the entire mixture is permitted only for priests but remains forbidden to laymen. The reasoning is as follows: the initial mixture contains 99 parts of ordinary fruit and one part of *terumah*. Since the ratio of *terumah* is greater than 1/101, the entire mixture is prohibited to laymen. With the introduction of 1/2 *seah* of *orlah*, the total amount is now 100 1/2, of which 99 is ordinary fruit, 1 is *terumah*, and 1/2 is *orlah*. The probability that something other than *orlah* will be drawn is the sum of the probability of drawing ordinary fruits and that of drawing *terumah*, to wit,

$$\frac{198}{201} + \frac{2}{201} = \frac{200}{201}.$$

Therefore, the *orlah* prohibition is nullified.

Also where exact relative frequencies are not available, addition is carried out in the Talmud. Our example is drawn from a protracted discussion of aspects of the law of levirate marriage. Implicitly, multiplication is involved as well. For the moment, though, we consider the aspect of addition only.

What is the probability that a pregnant woman will bear a live male child? This must be less than one-half. For, 'A minority [of pregnant women] miscarry, and of all live births – half are male and half female. Add the minority of those who miscarry to the half who bear females and the males are in a minority.'[5]

3 *M. Orlah* II 2

4 *J. Talmud, J. Orlah* II 2

5 *Yevamot* 119a. It is noteworthy that the Jerusalem Talmud treats this case as an instance of the double doubt, i.e. by Multiplication (see § 4.3 below), rather than as in the Babylonian Talmud by addition. 'There are two doubts: it is a doubt whether it is male or female and it is a doubt whether it is a viable child or not' (J. *Yevamot* XVI 1). Unlike the parallel passage in the Babylonian Talmud which refers to live births being equally divided in the Jerusalem text it is implicitly assumed that all foetuses are half males and half females.

4.2 MULTIPLICATION: CONDITIONAL PROBABILITIES

Multiplication arises naturally in two different ways. The first occurs in problems dealing with mixtures: 'One *seah* [a dry measure] of *terumah* fell into less than a hundred [of ordinary grain], and later [more] ordinary [grain] fell into the [mixture to make up 100:1 of the ordinary as against the *terumah*] ... it is permitted.'[6]

This early case illustrates a rudimentary kind of multiplication. Let the amount of ordinary grain in the initial mixture be x and in the final mixture be y. The rabbis thought in terms of relative frequencies, but but we might just as well use the modern terminology in which probability means relative frequency. The probability of drawing *terumah* T from the initial mixture M, written $p_M(T)$ is given by

$$p_M(T) = 1/(x + 1),$$

while the probability $p_F(M)$ of drawing an element of the initial mixture M from the final mixture F is given by

$$p_F(M) = (x + 1)/(y + 1),$$

and the probability $p_F(T)$ of drawing *terumah* from the final mixture is

$$p_F(T) = 1/(y + 1) = 1/(x + 1) \cdot (x + 1)/(y + 1),$$

so that

$$p_F(T) = p_M(T) \cdot p_F(M). \tag{1}$$

That this line of reasoning was in fact followed by the Talmudic rabbis is immediately apparent from the following far-reaching extension of the above ruling, which the early commentators illustrate by actual computations: 'One *seah* of *terumah* fell into less than a hundred [of ordinary grain] and thus [the mixture] became prohibited because of the doubt. Now, some of this doubtful mixture fell into another place [with an arbitrary amount of ordinary grain] ... it causes a state of doubt only in proportion.'[7] Not

6 *M. Terumot* v 9
7 *M. Terumot* v 6. See also *Terumah* 12a

only do we have here a general rule for computing the probability of drawing *terumah* from the final mixture or, more precisely, for calculating the relative frequency of *terumah* in the final mixture. Here another principle of inference is assumed, namely, that the frequency of *terumah* in a part or a sample from the first mixture is the same as in the entire mixture. This will be discussed in detail later. For the present, our concern is the multiplication rule only.

A special case of the multiplication formula phrased for acceptance rules is stated in the Jerusalem Talmud as follows: 'R. Z'eira in the name of R. Pedayah said: certain *terumah* prohibits up to a hundred [times as much ordinary fruits], and doubtful [*terumah*] up to fifty.'[8] The statement occurs in a context dealing with single figs, and is explained by a recent commentator: 'a doubt is valued at one half since both sides of the doubt are balanced. Therefore, in quality it is valued at 1/2, which is just 1/2 of its quantity.'[9]

The problem here is that on the one hand one single fig certainly cannot be considered 1/2 *terumah* and 1/2 secular; on the other hand, according to the strict multiplication rule, if the probability of the doubtful fig being *terumah* is 1/2, and it is mixed up with fifty others, we obtain

$$p_F(T) = 1/2 \times 1/51 = 1/102.$$

Strictly speaking, then, doubtful *terumah* prohibits only up to 49 1/2 times as much. However, when dealing with whole fruits, the acceptance rule must refer to integers rather than fractions, and the least integer exceeding 49 1/2 is 50.

Thus, we see that the rabbis multiplied together two prob-

8 *J. Terumot* IV 7

9 Rabbi Joseph Engel, *Gilyonei HaShas* to J. *Terumot* IV 7. See, however, Maimonides, *Mishneh Torah, Terumot* XIV 8–9, who apparently understands this passage differently. He rules, seemingly basing himself on this source, that neutralization of one part *terumah* requires 100 parts secular fruits only as long as the entire mixture is extant; but even if the ratio is 1:51 and only one was lost, the remainder is permitted. See also the gloss of Rabbi Abraham ben David and the commentaries of Rabbi Joseph Corcos and Rabbi Joseph Caro (*Kessef Mishneh*) on Maimonides.

abilities to obtain a product which is itself a probability. However, the factors are not just any probabilities. Rather they are related in a special way: in our example, the *terumah* in the final mixture comes exclusively from the initial mixture. The significance of this limitation will now be explained in terms of the modern theory.

If we let T stand for the set of all elements which are *terumah*, and M stand for the set of all elements in the initial mixture, then the set of *terumah* in the final mixture must come from among just those elements which are in both M and in T, namely those elements of the initial mixture which are *terumah*. This may be represented as the intersection of the sets M and T, written $M \cap T$. We may therefore write $p_F(M \cap T)$ instead of $p_F(T)$ in the multiplication formula above obtaining,

$$p_F(M \cap T) = p_M(T) \cdot p_F(M) \tag{2}$$

If one thinks of M and T as arbitrary properties, the product $p_F(M \cap T)$ can be understood to represent the relative frequency of elements having both properties M and T in the reference class F. This probability is given by the formula as the product of two probabilities. Now one of these factors, $p_F(M)$, is just the relative frequency of elements having property M in the reference class F.

If M and T are arbitrary properties, the symmetry of the situation suggests that the following be true.

$$p_F(M \cap T) = p_F(T) \cdot p_F(M), \tag{3}$$

where the reference class for all these probabilities is F. However, that is not the case in our formula. This is due to the special asymmetrical conditions which obtain in our example. Whereas there are elements of M in F that are not in T, there are no elements of T which are not also in M, i.e., T depends upon M. Therefore, $p_F(M \cap T) = p_F(T)$, and so formula (2) could be identical with formula (3) only if M is all of F making $p_F(M) = 1$. In that trivial case there would really be no multiplication at all. In modern probability theory, the factor $p_M(T)$ is called a 'conditional probability' and is usually written $p(T|M)$. It is the probability of T given a restriction to the set M.

Since in our example T is dependent upon M, $p_F(T) \neq p_M(T)$. Two properties M and T are said to be 'independent' if the conditional probability of one given the other is the same as its unconditional probability.

4.3 MULTIPLICATION: INDEPENDENT PROBABILITIES

Multiplication of independent probabilities also arises in the Talmud and is illustrated in the following instance. An adulteress is forbidden to live with her husband, so that he may divorce her without penalty. But this is not applicable if he claims that his bride committed adultery before the consummation of the marriage, even though it be established that she is not a virgin: 'It is a double doubt. It is a doubt: whether under him [i.e., during the period – usually one year – between formal betrothal and consummation] or not under him [i.e., prior to the betrothal]. And if you say that it was under him there is the doubt whether it was by violence or by her free will.'[10]

It is assumed that the chances are at most even that the incident occurred during the period of betrothal. But even if such were the case, since a woman violated by another man does not become forbidden to her husband, it can be argued that there is a chance that she did not submit of her own free will. One of the early commentators points out that rape is known to occur infrequently. This is countered by the argument that even so the probability of adultery is still less than half in view of the first doubt. Thus, half such cases occur 'under him' and of these some are rape, leaving only a minority of culpable adultery.[11]

Rabbi Bezalel Ashkenazi (sixteenth century) argues against this kind of computation when dealing with independent events. He distinguishes this case from straightforward addition.

When the minority who abort is added to the 1/2 of females, both reduce the strength [numbers] of males by virtue of the same cause [both are not males]. The minority of rape and 1/2 'not under him' as arguments for

10 *Ketubot* 9a

11 Tosafot to *Ketubot* 9a s.v. ואי. See Rabbi Jacob Joshua Falk, *Penei Yehoshua*, on this text

leniency do not stem from the same cause. So each one alone is opposed to 'under him willingly' – the minority of rape as against 'willingly,' and the 1/2 'not under him' against 'under him.' Since this is so, how is [the frequency of] 'willingly' reduced by adjoining the minority of rape together with 'not under him,' and how is 'under him' reduced if 'not under him' is joined with the minority of rape? Each stands alone – the majority of 'willingly' opposes the minority of rape which is therefore annulled, and there remains only the even doubt whether 'under him' or not.[12]

He seems to be saying that on frequency considerations alone we have no way of knowing that any of the half who had illicit relations during the period of betrothal did so under duress. Perhaps the minority of rape is to be found exclusively among the other half. Or, perhaps there is some other distribution. Since we know nothing at all about how these two considerations may be related, a decision on each question must of necessity be independent. Thus for an arbitrary woman, 'follow the majority' tells us that rape was not involved. That is a conclusion warranted by the evidence. As to the question of when the incident occurred, since the doubt is half and half, it cannot be resolved. Only in the case where both doubts are half and half do we have a genuine double doubt. For, if we knew that the frequency of rape is 1/2 and the frequency of 'under him' is also 1/2, then no decision on either question would be possible and a double doubt would exist which warrants leniency.[13]

This is indeed a fundamental objection to multiplication of independent probabilities if they are conceived of in empirical frequency terms alone. It is a criticism that applies to modern frequency interpretations of probability as well.[14] To make it general, the point is simply this. If two attributes A and B or their negations can be pre-

12 Rabbi Bezalel Ashkenazi, *Shittah Mekubetzet* to *Ketubot* 9a (folio 35d)

13 Strictly speaking, it is not inconsistent with the data to suppose that all cases 'under him' were willing and all cases 'not under him' were rape. However, to assume that such is actually the case is far-fetched. If even only some of the 1/2 cases 'under him' are also rape, leniency is justified. This seems to be implicit in the argument.

14 See J.M. Keynes, *A Treatise on Probability*, p.106

dicated of all the members of a given set X, and the proportions having the attributes A and B are $p(A)$ and $p(B)$ respectively, then it is possible for the proportion $p(A \& B)$ having both A and B to vary within wide limits. For the set of elements having A, namely X_A, to have no members in common with X_B, it is necessary but not sufficient that

$$p(A) + p(B) \leqslant 1.$$

If this condition is satisfied, it is conceivable that there are no elements having both A and B and so $p(A \& B) = 0$. Otherwise, a necessary minimum for $p(A \& B)$ is given by

$$\min p(A \& B) = p(A) + p(B) - 1.$$

Clearly

$$\max p(A \& B) = \min (p(A), p(B)),$$

since there cannot be more elements having both attributes than there are having either attribute alone.

Between these two extremes any proportion of elements having both A and B is admissible, since it is consistent with the given data. Taking probabilities to refer to actually observed relative frequencies, nothing more can be said about $p(A \& B)$, in the absence of further evidence. For who is to say what is in fact the actual proportion of elements having both properties A and B? That proportion certainly cannot be inferred from the fact that, as far as is known, A and B are not related in any necessary way. If probabilities are to be based on empirical frequencies when dealing with one character, the same must hold for combinations of two or more characters. Only if one actually knows the distribution of elements having both characters is one justified in making inferences with respect to such elements.

We shall see later that the above text on the double doubt was understood by other commentators in a different way, involving another kind of probability conception.

4.4 AXIOMS OF PROBABILITY THEORY

The modern theory defines the non-negative probability function for infinite as well as finite sets. In the frequency interpretations of von Mises[15] and Reichenbach[16] infinite reference classes are fundamental, although in other interpretations finite classes are used.[17] Given a certain structure of these sets, the special properties of the probability function can be specified in three axioms from which, together with the usual axioms of mathematics, the entire mathematical theory is derivable.[18]

A1 *Axiom of total probability*

If F is the reference class, $p_F(F) = 1$. This means, for example, that if all the balls in an urn are white, the probability of drawing a white ball is 1, namely, a certainty.

A2 Addition axiom

If sets T and R have no elements in common, and $T + R$ represents their union, that is to say the set of all elements belonging either to T or to R, then

$$p(T + R) = p(T) + p(R).$$

We can use a rabbinic illustration mentioned above. Suppose T is the set of *terumah* and R the set of secular fruits. The reference class is the final mixture. No element of *terumah* is also secular nor is any of the secular *terumah*. Since the two properties are mutually exclusive, it follows therefore that the probability of drawing either *terumah* or ordinary fruit is given by adding the individual probabilities.

A3 *Multiplication axiom*

$$p_F(M \cap T) = p_F(M) \times p(T|M).$$

15 Richard von Mises, *Probability, Statistics and Truth* (London, 1939)

16 Hans Reichenbach, *The Theory of Probability* (University of California Press, 1949)

17 See, for example, Bertrand Russell, *Human Knowledge* (London, 1948) Part v, pp. 353–418; R.B. Braithwaite, *Scientific Explanation* (Cambridge, 1964), pp.191 ff.

18 A.N. Kolmogorov, *Foundations of the Theory of Probability* (Chelsea, 1950), p.2. The version presented here follows more closely that given in Henry E. Kyburg, *Probability and Inductive Logic* (Macmillan, 1970), pp.15–18.

This gives the probability of the set of elements having the properties of both M and T. As explained above, if M and T are independent, the second factor on the right becomes $p_F(T)$. Otherwise, it is the conditional probability of T restricted to the set M.

4.5 INVERSE PROBABILITIES AND BAYES'S THEOREM

An important theorem which, since it was published in 1763, has been the centre of much philosophical controversy is the celebrated Bayes's Theorem.[19] Its importance stems from the fact that it provides a means for computing the conditional probabilities of competing hypotheses relative to given evidence.

The theorem is easily derived from the multiplication axiom A3. To indicate the matter under discussion let us change the letters T and M, writing in their place H and E for hypothesis and evidence, respectively, and since there is no possibility of confusion, the subscript F for the reference class may be omitted. Thus A3 takes the form

$$p(H \cap E) = p(H) \times p(E \mid H). \quad (3)$$

Also

$$p(E \cap H) = p(E) \times p(H \mid E). \quad (4)$$

Since $H \cap E = E \cap H$, both representing the set of elements common to both sets E and H, the left-hand sides of equations (3) and (4) are equal. Consequently, the same is true for the right-hand sides.

$$p(E) \times p(H \mid E) = p(H) \times p(E \mid H) \quad (5)$$

$$p(H \mid E) = [p(H) \times p(E \mid H)]/p(E). \quad (6)$$

This is Bayes's Theorem and it gives the conditional probability of H on the basis of E, in terms of the prior probabilities of H and E, as well as the probability of observing E if H is true.

Suppose there are two hypotheses H_1 and H_2 such that either

19 Thomas Bayes, 'An essay towards solving a problem in the doctrine of chances,' *Philosophical Transactions*, vol.53 (1763), 392

H_1 or H_2 must be true but both together cannot be. As an illustration, let us consider a case raised in the Talmud[20] about a widow who married her brother-in-law before the required three-month waiting period had elapsed, and although at three months after her husband's death there were no external signs of pregnancy, she gave birth to a child scarcely six months later, i.e., hardly nine months after her first husband's death. The rabbis propose two exclusive and exhaustive alternatives. H_1 is the hypothesis that the child is a full-term baby belonging to the first husband. H_2 is that the child is a seven-month baby fathered by the second husband.

Now, 'most women bear [a child] only after nine months' pregnancy.' This gives the prior probability of the hypothesis that the child is a full term one, $p(H_1)$. On the other hand 'for most women who deliver at nine months, pregnancy is already recognizable at one third term,' whereas the evidence in this case was negative. What is the probability of an absence of external signs of pregnancy $p(E)$? Since one of either H_1 or H_2 must be true, we compute the probability of the evidence on each hypothesis. Suppose H_1, then $p(E|H_1)$ is a minority, for only a minority of pregnancies that last nine months are still concealed at three months. On the other hand, if H_2 is true, then $p(E|H_2) = 1$, since at three months after her first husband's death, the pregnancy would not have been more than one month old and this would surely not be recognizable. The total probability $p(E)$ is obtained from

$$p(E) = (p(H_1) \times p(E|H_1)) + (p(H_2) \times p(E|H_2)). \tag{7}$$

Substituting in equation (6) for Bayes's Theorem gives

$$p(H_1|E) = \frac{p(H_1) \times p(E|H_1)}{(p(H_1) \times p(E|H_1)) + (p(H_2) \times p(E|H_2))} \tag{8}$$

and a corresponding formula for $p(H_2|E)$.

Now, the Talmud is dealing here with unenumerated populations. We have already seen that for the rabbis 'standard' majorities range somewhere from less than 2/3 to more than 3/4.[21]

20 *Yevamot* 37a
21 See above pp.41 and 44

Therefore, we can attempt an approximate calculation of the posterior probabilities of the hypothesis H_1 and H_2 given the evidence E. Let us use the 2/3 figure first. Then

$$p(H_1) = 2/3, p(H_2) = 1/3;$$
$$p(E|H_1) = 1/3, p(E|H_2) = 1.$$

From formula (8), we obtain then,

$$p(H_1|E) = \frac{\frac{2}{3} \times \frac{1}{3}}{(\frac{2}{3} \times \frac{1}{3}) + (\frac{1}{3} \times 1)} = \frac{2}{5}$$

and $p(H_2|E) = 3/5$.

Thus, the majority of 2/3 which is the prior probability of H_1 is transformed by the evidence into a minority of 2/5.

However, if we take as an approximation of an unenumerated majority the value 3/4, it follows that

$$p(H_1|E) = 3/7, p(H_2|E) = 4/7.$$

On the other hand, if we consider the following approximation,

$$p(H_1) = 3/4, p(H_2) = 1/4;$$
$$p(E|H_1)\ 1/3, p(E|H_2) = 1,$$

the results are

$$p(H_1|E) = 1/2, p(H_2|E) = 1/2.$$

This explains the Talmudic argument: 'The majority of women bear [a child] only after nine months' pregnancy ... and for most women who bear at nine months, pregnancy is already recognizable at one third term, but this one, since it was unrecognized at one third term, the majority is weakened.' Rashi adds: 'You cannot put her into the majority of women, but it is a doubt whether she belongs to the minority or the majority.'

Of course, this is not to suggest that the rabbis had Bayes's Theorem as such. Yet, it is clear that they did have a concept of prior and posterior probability and they recognized the nature of the relation between them.

4.6 SUMMARY

It has been shown that in the limited context of a finite frequency interpretation, and a probability function defined over a finite range only, the Talmud operated with the fundamental axioms of probability. The examples cited in this as well as in the previous chapter demonstrate that the rabbis had not only the notion of acceptance rules which amounts to a definition of the probability function, but they had also the rudiments of the calculus of probability. Moreover, we have found too an awareness of the philosophical difficulties inherent in the multiplication axiom, difficulties which trouble modern frequency theorists as well.

5
Logical alternatives

5.0 THE CLASSICAL DEFINITION OF PROBABILITY

In his pioneering treatise *Doctrine of Chances,* which appeared in 1718, Abraham De Moivre[1] gave a definition of probability which has the advantage of enabling probabilities to be evaluated a priori, thus obviating the need for empirical vindication of the addition and multiplication axioms. Today it is often referred to as the classical or Laplace definition of probability. Unlike the frequency interpretation which requires a knowledge of the actual distribution in a collection, the so-called classical definition allows one to assign probabilities even to single events. What is the probability of getting a two on a single toss of a die? Calling the case of a two 'favourable,' the numerical value of its probability is the quotient obtained on dividing the number of favourable possible cases by the total number of all equally possible cases. For a true die, there are six outcomes, all of which are equally possible, but only one is favourable; therefore the probability of throwing a two is 1/6.

In order to apply the classical definition, one must find a way of counting all the alternative possibilities. However, that is not

1 '... If we constitute a Fraction whereof the Numerator be the number of Chances whereby an Event may happen, and the Denominator the number of all the Chances whereby it may either happen or fail, that Fraction will be a proper designation of the Probability of happening.' (2nd ed., p.1)

enough. It is required also that these alternatives may legitimately be presumed to be equally possible. Given a new die that has never yet been cast, is it reasonable to suppose that all sides may turn up with equal ease? Of course, if it happens to be known that the die is loaded, such a supposition is patently false. But if there is no reason to suspect bias, is it not a rationally defensible assumption that all the possibilities are equally likely?

The postulate to which one appeals in order to justify such an assumption is that two possibilities are equiprobable if and only if there is no cause to suppose otherwise. This was long known as the Principle of Non-Sufficient Reason and, according to Keynes,[2] was introduced by Jacob Bernoulli.[3] Keynes himself renamed it the Principle of Indifference, and in order to avoid the paradoxes that the principle had spawned he specified restrictions on it. One such qualification is: 'The Principle of Indifference is not applicable to a pair of alternatives, if we know that either of them is capable of being further split up into a pair of possible but incompatible alternatives of the same form as the original pair.'[4]

With the aid of the Principle of Indifference, the axiom of multiplication can be proved. This is accomplished somewhat as follows. Given two sets A_1 and A_2 such that A_1 consists of m_1 white and n_1 black elements, while A_2 consists of m_2 white and n_2 black ones, the product set $A_1 \cdot A_2$ is defined as the set of all possible pairs consisting of one element from A_1 and another from A_2. To each element of A_1 one can match every one of the $(m_2 + n_2)$ elements of A_2. In all, there are then $(m_1 + n_1) \cdot (m_2 + n_2)$ such possible pairs, and the Principle of Indifference allows us to assume that they are equally probable. If we ask what is the probability in $A_1 \cdot A_2$ of the set of exclusively white pairs, the answer is

$$\frac{m_1 m_2}{(m_1 + n_1)(m_2 + n_2)}.$$

The number of all possible combinations of m_1 whites with m_2

2 J.M. Keynes, *A Treatise on Probability*, p.41
3 *Ars Conjectandi*, p.224
4 *A Treatise on Probability*, p.61

whites is simply the product $m_1 m_2$, and the quotient of this product divided by the number of all pairs in $A_1 \cdot A_2$ gives the sought-for probability. But this is exactly the product of the probabilities of the sets of whites in each of A_1, A_2. In symbols,

$$\frac{m_1 m_2}{(m_1 + n_1)(m_2 + n_2)} = \left(\frac{m_1}{m_1 + n_1}\right)\left(\frac{m_2}{m_2 + n_2}\right),$$

which is what the multiplication axiom requires.

Thus is solved what Popper has called 'The Fundamental Problem of the Theory of Chance.'[5] Chance, which seems to be characterized by utter incalculability, is brought within the purview of a calculus and, instead of complete unpredictability, one obtains laws of chance.

We shall see that a probability conception based on logical alternatives is implicit in some very early rabbinic sources. The explicit identification of equipossible with equiprobable occurs no later than the twelfth century, and implicit use of the concept as early as the second century. The rabbis were troubled too by the difficulty of defining ultimate alternatives. Furthermore, some questioned the assumption that all possibilities are equiprobable. There were those who sought some sort of physical ground for equiprobability much like the recent view that argues for probability as measuring a propensity in the physical system. Others groped towards a definition of randomness, which would provide a criterion for determining when alternatives may be regarded as being equally likely.

5.1 ENUMERATING THE ALTERNATIVES

Legal questions often arise as a result of unique occurrences and there is no class of similar events which can yield majorities relevant to the issue in doubt. For instance: 'Two bins, one containing *terumah* and one ordinary [grain], and a bushel of *terumah* fell into one of them, but it is not known into which one.' Although there is no reference class of similar cases from which one might derive

5 K.R. Popper, *The Logic of Scientific Discovery*, p.150

empirical probabilities, yet one can readily conceive of all the possibilities that might occur in a case like this. There are only two choices – the *terumah* fell into either one or the other container, and assuming that both were equally accessible, both seem equally likely. Even where there is no evidence to indicate a tendency towards one alternative rather than another, and on the face of it, the alternatives are equally probable, the principle "Follow the majority" may still find application. Rabbi Joseph di Trani the Elder (sixteenth century) explains how. Attention is now focused on the number of possible alternatives. If the law is lenient for some alternatives but stringent for the others, this gives two sets of possibilities, and the decision follows the larger set. In other words, follow the majority of possibilities.[6]

Such a case is dealt with in the Tosefta. We recall that *terumah* is sacred, so that when mixed with ordinary grain it renders the whole forbidden to all except priests: 'There were two bins – one of wheat and one of barley [the amount of each being unknown] and besides them two bushels of *terumah* – one of wheat and one of barley. [The contents of] One of the bushels of *terumah* fell into [another one of the containers] and the other [bushel] was lost – both bins are permitted.'[7]

What are the possibilities in this case? The wheat *terumah* could have fallen into the barley *terumah* which was then lost, or it might have ended up in the bin of ordinary wheat rendering it forbidden, while the barley *terumah* was lost. The third possibility that the wheat *terumah* fell into the barley bin obviously was not feasible since it could be detected easily. There are then only two alternatives for the wheat *terumah*, and similarly two for the barley *terumah*, for a total of four equally likely cases. Of these, in two cases, all the *terumah* was lost and the status of the ordinary grain is unaffected. Thus each of the wheat and barley bins has only one chance in four of having *terumah* mixed in, and is therefore permitted.

An interesting instance of counting alternatives occurs in a pas-

6 Rabbi Joseph di Trani (the Elder), *Responsa*, vol.2, *Yoreh Deah*, no.2 (6a, col.2)

7 *T. Terumot* VI 14. The explanation of this text follows Lieberman's commentary, p.392.

sage in Tosafot.[8] The argument runs thus. Assume that it is impossible to determine precisely that two lengths *a* and *b* are equal, since any measurement is subject to error. Moreover, it is an infrequent occurrence for two lengths to be equal. Nevertheless, when they appear to be equal, we may suppose it more likely that *a* is greater than or equal to *b* than that *a* is less than *b*. Similarly, the probability is greater that *a* is less than or equal to *b* than that *a* is greater than *b*. For there are three possible cases: (1) *a* is greater than *b*, (2) *a* is less than *b*, and (3) *a* equals *b*. It follows that the likelihood of cases 1 and 3 together is greater than that of case 2 alone, and similarly for cases 2 and 3 as against case 1 alone.

Although it is explained that case 3 has a low probability relative to either of the other two cases, the fact is not noted that cases 1 and 2 represent infinite collections while equality is only one point, and that would vitiate the argument. It is plausible that the writer of the gloss took it for granted that for a given *b* in practice one can construct only a finite number of lengths *a* close to but less than *b*, and a similar number greater than *b* but close to it.

In any but the very simplest problems, the enumeration of the number of possible cases involves the computation of combinations and permutations. Although this branch of arithmetic was developed early among the Jews, as we shall see later, I have not been able to find explicit computations with reference to probability. In all the rabbinic examples, of which those given here are typical, all the possibilities can be enumerated without difficulty. Nevertheless, it is worthwhile to take a look at another couple of problems based on the idea that all possible permutations are equally likely, for they evoked some fundamental insights.

5.2 COMBINATIONS *VS* PERMUTATIONS

We have seen that the Talmud assumes that for a live birth the chances that the child will be a boy or a girl are even,[9] for each birth is an independent event. Suppose one or more women have given

8 *Sukkah* 15a and b s.v. פרו״ן
9 See above p. 50

birth to a number of children. The order of birth, how many children each mother bore, which child belongs to each, are all unknown. What then is the probability that a particular woman bore boys and girls in a specified sequence? In this case the births are not to be considered as independent events. Assuming that every permutation of boys and girls, in the given number, is equally likely, one can compute the probability of any particular sequence.

This question arises in connection with the biblical command: 'You shall set apart to the Lord all that first open the womb. All the firstlings of your cattle that are males shall be the Lord's. Every firstling of an ass you shall redeem with a lamb ... Every first-born of man among your sons you shall redeem.'[10] The price of redemption for human children is five shekels.[11]

The Mishnah rules in a hypothetical case which is parallel to similar cases referring to cattle, where it is not unusual for offspring to be mixed up:

> If one woman had already borne young and another had not before given birth – [these two being the wives] of two men – and they bore two males, he whose wife had not before borne a child must give five pieces of silver to the priest; a male and a female – the priest gets nothing [since the probability is only 1/2 that the mother giving birth for the first time had the boy] ...[12]

Maimonides, in his Code, adds another case discussed in the Gemara[13] and explains it as follows:

> If they bore two males and a female, the husband of the primipare must give five pieces of silver, because he would be free in only two cases [1. if his wife bore the daughter only, and 2. if she bore the daughter first followed by a son] while if his wife bore only male offspring, [i.e., two cases for one male child and one case for both male children] he is obligated, and if she bore a male and a female – he is obligated as well, unless the

10 Exodus 13:12–13
11 Numbers 3:47
12 *M. Bekhorot* VIII 6
13 *Bekhorot* 49a

female was first. Since the chance is remote, [two cases vs. four], he must pay redemption.[14]

Since the mothers are wives of different men, if one of the boys is a first-born, his father must be the husband of the primipare. There is then an independent method of identifying the children other than through the mother, namely through the father. Of course, in practice it does not work, since we cannot distinguish which is which. However, if the father says: 'If my son is the first-born of his mother, let this be his redemption,' he is referring to one particular boy – that one who is his son. If both are his sons, then there is no question but that he is obligated to pay redemption. Therefore, all we need to decide is whether the first-born of the primipare was a son, but we need not determine whether a particular boy is the first-born. It suffices to count all possible ways of dividing the three children among the two mothers, as quoted above. Consequently, we count as only one possibility the case when she had both boys. Similarly, if she had only one boy, we count as only two possibilities the cases when the other woman had a boy first followed by a girl or vice versa.

However, if the two women are both wives of the same husband, the issue to be resolved is somewhat different. Although, curiously, Maimonides does not say so explicitly, he seems to base himself upon the rule that the obligation of the father to pay redemption to the priest is primarily a duty to his son. When we know that both boys belong to the same father, but not necessarily the same mother, we need to ascertain, for *each* boy, whether it is likely that he is a first-born. In dealing with this question, each case must be divided into two according as one or the other of the boys belongs in a specified place. Thus to each arrangement there corresponds a different permutation involving two boys, yielding

14 *Mishneh Torah, Bikkurim* XI 30. I have completed the enumeration of all the cases (in square brackets), although Maimonides himself mentions only some of them here. However, a little further on, in another instance, Maimonides explicitly lists these cases (*Bikkurim* XII 21). That passage is quoted below, p. 69.

twelve permutations in all, of which in four cases boy *A* is a first-born and in four cases boy *B* is, while in the remaining four cases the girl is the first-born. Since for each boy the chances are no greater that he is first-born than is the girl, Maimonides decides that there is no obligation of redemption when both women are married to the same man.[15]

5.3 ULTIMATE ALTERNATIVES

In the case of the male first-born of a donkey, a prohibition applies to an unredeemed animal. Redemption is accomplished by giving a lamb to the priest. Even when the priest has no claim because the status of the animal is in doubt, if the chances are even, a symbolic rite of redemption is required to lift the prohibition. Because the possibility that the animal is a first-born is no less likely than the alternative, the ritual of exchange for a lamb must be performed, although the lamb need not be handed over to the priest. This is so if a primiparous ass bore a male and a female and it is not known which was first, or if two she-asses, one primipare and one not, bore a male and a female and it is not known which belongs to which.

The Mishnah speaks of another case of one who had two she-asses who gave birth for the first time, and Maimonides interprets it in this way: 'If they bore two females and a male ... the priest gets nothing and even the ritual of exchange for a lamb is not required, because there are here many doubts – perhaps the one bore a male and the second bore two females, or perhaps she bore a female and the other a male followed by a female or a female followed by a male.'[16]

It seems best to present the data in tabular form. Let *m*, *f* stand for male and female respectively; columns I and II list the offspring of the first and second mothers respectively. Maimonides itemizes three possibilities if mother I had one colt only.

15 Ibid. XI 25
16 *Mishneh Torah: Bikkurim* XII 21

Table A

	I	II
Case 1	*m*	*f*, *f*
Case 2	*f*	*m*, *f*
Case 3	*f*	*f*, *m*

Table B

	I	II
Case 1′	*f*, *f*	*m*
Case 2′	*m*, *f*	*f*
Case 3′	*f*, *m*	*f*

He leaves it to his readers to see that if mother I had two colts the columns in Table A are reversed, as in Table B.

Apparently Maimonides sees the obligation of redemption for animals as rooted in the ownership of the mother. Thus the question to be decided is: did this she-ass bear a male first-born?[17] Looking at column I in both Table A and Table B, in two out of three cases the first-born is female. Therefore, there is no obligation of redemption. The same is true for the second mother, as can be seen in column II.

However, Rabbi Abraham ben David of Posquieres, in a curt gloss on Maimonides' text is sharply critical. He says, 'There is a misconception here. The doubt is [only] whether the male followed the female.' This criticism is taken up by subsequent writers[18] who explain that the real issue is whether the male is a first-born or not. Now there are six cases in all, and it is clear that the male is first-born in four out of the six possible arrangements, namely cases 1, 2, 1′, and 2′. There is an alternative explanation according to which the doubt is an even one, for the male either followed a female or it did not, and there are no other possibilities.[19] That this is what Rabbi Abraham had in mind is more likely since he too does not require payment to the priest, which he would have done if he considered it 2:1 that the male is first-born.

This difference of opinion is but one example of a very troublesome question that was raised quite early in connection with the definition of alternatives. How are they to be defined and with

17 Rabbi Joshua Falk Kohen, *Drishah* to *Tur: Yoreh Deah* 321, n.4

18 Rabbi Judah Rosanes, *Mishneh LaMelekh* to the cited text of Maimonides. Note, however, Maimonides' own remark, *Mishneh Torah: Bekhorot* V 2

19 Commentary of Rabbi David Ben Zimra on *Mishneh Torah*

reference to which criteria are we to decide what is a simple alternative which counts as only one, and what a compound one that counts for more than one?

The matter was discussed mainly in connection with the 'double doubt.' We have already come across the 'double doubt' as seen from the point of view of relative frequencies. The simplest instance of the distinction between simple and compound alternatives occurs in a double doubt mentioned by Rabbi Shlomo ben Adret,[20] namely, if one forbidden article is mixed up with a similar permitted one, and one of these two is mixed up with a third similar and permitted one. For each article in the final pair there are two equally likely alternatives, either it is the permitted one that fell in last or else it stems from the first mixture. But the second alternative itself consists of two subcases, each of which is equiprobable.

In dealing only with single objects, there is no assumption required about the sample distribution such as is the case in more numerous mixtures, and moreover, the alternatives are clear and well defined – one simple and the other compound. The discussions in the sources on the double doubt generally revolve around more complex issues, and often there is contention as to whether the alternatives can be regarded as equally probable at all.

Thus to go back to the case of the woman accused of adultery after betrothal, some commentators reject the frequency interpretation altogether and treat the matter as if it were a case of simple alternatives; 'it is a doubt whether by violence [*V*] or by choice [not-*V*], a doubt whether after betrothal [*A*] or before [not-*A*].'[21] There are then four different combinations of the alternatives, three of which are in her favour (*V* & *A*, *V* & not-*A*, not-*V* & not-*A*), and only one against her (not-*V* & *A*).

In a remark of the Tosafot,[22] the suggestion is advanced that for a bride that was betrothed very early, even if the incident occurred after betrothal, there is still another doubt – perhaps it was when she was very young, and the law provides that seduction

20 *Torat HaBayit HaArokh* IV 2 (74b)

21 *J. Ketubot* I 1. See above p. 54

22 *Ketubot* 9a, s.v. ואי

of a juvenile is to be considered as violence. In rejecting this idea, it is argued that 'the category of violence is one,' the reasoning being that in the dichotomy violence/choice, there is no room for a detailed analysis of degrees of motivation, since for legal purposes in the given context there are only two possible states. Therefore, each of the alternatives must be regarded as simple and whatever is legally 'violence' cannot be subdivided further. The phrase 'the category of violence is one' entered into the legal jargon of the subsequent authors and is used as a technical term for rejecting a spurious subdivision of a legally simple possibility. But clearly what is legally simple depends upon what legal criteria are relevant. As we have seen in the case of the first-born, the same case can often be referred to differing legal considerations and thus possibilities regarded as simple in one opinion are seen as compound in another.

Although many centuries separate us from the medieval rabbis and the context of the discussion is now philosophical rather than legal, the same differences over defining alternatives persist. Interestingly enough, a present-day writer makes the following suggestion: 'The character of the *principal alternatives* is to be determined by the intent of the question to which the prediction-statement is a proposed response.'[23]

5.4 ARE LOGICAL ALTERNATIVES EQUIPROBABLE?

A more basic consideration is whether, in the absence of evidence, one is justified at all in assuming that logical alternatives are equiprobable. The matter is discussed by Rabbi Isaac bar Sheshet (1326–1408) in a responsum concerning the double doubt.

[When the Talmud says that] half of all [children born] are male and half female, it is certain and necessary, for thus did the King of the Universe establish it for the preservation of the species. Therefore of necessity, of all pregnant women those who bear males are a minority, for some abort,

23 N. Rescher, 'On prediction and explanation,' *British Journal for the Philosophy of Science*, 8 (1958), 286

and this is inescapable. But, here we cannot say that of all those who have illicit relations – half do so after betrothal and half before ... for whence do we know that it is half and half? It is only that we say the matter is in doubt, for the one or the other is possible; and even if we add all the cases of rape together with those not under him, the willing adulteress after betrothal is not certainly in the minority, and we can still say the one or the other is possible ... and the doubt still exists.[24]

It is unacceptable to treat this as a question of equiprobable logical alternatives and then to combine probabilities so derived in one computation with probabilities based on frequencies. One might just as well describe the alternatives as either she was a willing adulteress or not 'and we can still say the one or the other is possible.'

Rabbi Isaac bar Sheshet strikes at the root of the problem presented by the Principle of Indifference – can sound knowledge be derived from ignorance?[25] Lest the difficulty be attributed to ambiguity in establishing the ultimate alternatives, he raises the matter of the distribution of males and females where that is not the issue. Although observation confirms that male and female are equiprobable, even that is not sufficient to warrant a prediction, unless there is positive reason why the observed distribution in a known population may be assumed to obtain in an unobserved population.

A priori probabilities must be grounded in some kind of natural necessity if they are to have any validity. For the equal frequencies of males and females such a cause exists, so it is not only the empirically known frequencies that provide the relevant probabilities. Rather, they are a consequence of a teleological law designed for 'the preservation of the species.' By bringing in a consideration of a law of nature at work, Rabbi Isaac bar Sheshet assures himself of certitude in evaluating the probabilities of boys and girls.

24 Rabbi Isaac bar Sheshet, *Responsa*, no.372

25 For a discussion of the nineteenth-century empiricist criticism of the Principle of Non-Sufficient Reason, as it was then called, see J.M. Keynes, *A Treatise on Probability*, pp.85 ff.

5.5 A PROPENSITY INTERPRETATION OF PROBABILITY

It seems that there was a school of thought which sought to ascribe probabilities to a disposition of matter to behave in different ways according to statistical norms. Maimonides in his summary of the Aristotelian system as adapted to his needs, explains that all change is the result of the exchange of forms.

Matter is the bearer of form and when one form is replaced by another, we observe it as change. Where a transformation can result in one of a number of alternatives, this means that there exist a number of forms, each of which can be borne by the matter of the object in question. Maimonides postulated that possibilities are inherent in the capacity or propensity of the matter to receive given forms.[26] Where logical alternatives are equipossible, then it would appear that this is due to the equal tendency of the matter to receive each of the relevant forms. However, it certainly cannot be argued that, of necessity, matter is equally disposed to receive all forms, or that any set of alternatives is equiprobable without further knowledge. Thus Maimonides explains elsewhere that 'among contingent things some are very likely, other possibilities are very remote, and yet others are intermediate.'[27]

Maimonides makes the fundamental point that only if there is equal propensity to the alternatives can they be said to be equally probable. In that case, he argues, all possibilities will, in fact, be

26 *Guide of the Perplexed*, Introduction to Part II, Propositions 23 and 24 (Pines, p.239). The concept that contingency denotes a disposition of matter is apparently due to Ibn Sina. See Even-Shmuel's commentary on the *Guide*, vol.3, 33, n.42. Even-Shmuel also quotes a responsum of Maimonides in which he refers to the distinction between potentiality and contingency (pp. 31–2). See also Israel Efros, *Philosophical Terms in the Moreh Nebukim* (Columbia University Press, 1924), p.11

According to Gilson, Maimonides' views on the nature of the contingent were taken up by the Latin schoolmen, especially Albertus Magnus and Thomas Aquinas (Etienne Gilson, *Le Thomisme* [Paris, 1923], pp.60–1)

27 Maimonides, *Sefer haMitzvot*, Negative Commandment 290. Maimonides observes that one must take account of both 'a possibility in the matter to become that particular thing, and a possibility in the agent to produce that particular thing' (*Guide* II 14, Pines, p.287); cf. below p.75, n.30.

realized in due time if chance alone is at work. If, in fact, one outcome predominates, we can conclude that some causative factor is operative. 'There can be giving of preponderance and particularisation only with respect to a particular existent that is equally receptive of two contraries or of two different things. Accordingly it can be said of that inasmuch as we have found it in a certain state and not in another, there is proof of the existence of an artificer possessing purpose.'[28]

Has he not, in fact, substituted one unknowable for another? If we cannot tell a priori whether the alternatives are equiprobable, how can we tell whether the propensity for them is equal? Maimonides avoids this dilemma, because he applies his principle in cases where the same kind of matter has already been observed to be equally receptive to both alternatives.

Such a concept provides a physical basis for probability under a frequency interpretation, no less than under the 'classical' one. While one can speak therefore of a 'dispositional' or 'propensity' interpretation of probability, it is not an alternative to the other two. Rather it is meant to erect an underpinning for those other interpretations.

A notion of 'dispositional probability' was advanced in the last century by Charles S. Peirce[29] and has been developed by Karl Popper,[30] who gave it a modern formulation. However, its basic weakness remains in that it obviously does not apply to most uses of probability theory. While it seems to offer a physical and objective ground for probability, as for example in quantum theory, it is far too vague and has no practical advantages over an empirical frequency interpretation. Even today, just as in antiquity, it is

28 *Guide* I 74 (Pines, p.220)

29 Charles S. Peirce, *Collected Papers* 2.664 '... a die ... has a property, quite analogous to any *habit* that a man might have.'

30 Karl R. Popper, 'The propensity interpretation of probability,' *British Journal for the Philosophy of Science,* 10 (1959), 25–42

Popper remarks: 'propensities exhibit a certain similarity to Aristotelian potentialities. But there is an important difference: they cannot, as Aristotle thought, be inherent in the individual *things* ... they are relational properties of the experimental arrangement' (p.37). To my knowledge, he does not refer to the medieval doctrines on contingency.

ultimately observation of frequencies that yields reliable probabilities.

5.6 WHAT IS 'RANDOM'?

Although there were among the rabbis those who argued against admitting alternatives as equally probable without sound reasons, nonetheless, no one questioned the propriety of considering all possible permutations as equiprobable in phenomena which are generally recognized as due to pure chance. Things falling into or being drawn from mixtures at random are treated in the earliest as well as the latest sources in this way, as we have already seen. The need was felt for a more precise characterization of 'chance' phenomena. This was not described, as is the case in modern times, as a search for a definition of randomness, but it amounted to much the same thing. Essentially the question asked was: Is there a general rule that enables one to say when alternatives may be treated as equiprobable?

In the Talmud, a variety of cases are dealt with in which the alternatives are regarded as equal without hesitation, and in attempting to characterize these types of cases, the early commentators formulated a rather general principle which, though not a definition of randomness, is of considerable power. Let us consider one example:

A house collapsed on a man and his wife. The husband's heirs [in the absence of common children] say: 'The woman died first and the husband afterwards [so that he inherited her and we inherit him]'; the woman's heirs say: 'The husband died first and the wife afterwards.' ... Bar Kappara taught: 'Since these are heirs and these are heirs, they divide [the inheritance between them]'[31]

In such cases and similar ones where all the claims may, in fact, be warranted, the Talmud rules 'property that is in doubt is divided

31 *Bava Batra* 158a ff. See the commentary of Rabbi Samuel ben Meir on this passage

equally' among all the claimants. Rabbi Mordecai ben Hillel[32] analyses some of these instances and explains that the rule applies 'when the doubt remains ... forever, for it is such as cannot ever be resolved.' On the other hand, it is not appropriate in 'a matter which can be resolved.' Where ignorance is remediable, the fact that no decisive evidence is available at present merely imposes the duty to search for it diligently.

5.7 UNKNOWABILITY AND INDETERMINACY

Another approach to the unknowable arose out of a notion that appears in the rabbinic writings in an attempt to offer a physical ground for probability, namely, that of the confluence of various physical forces or laws. This probably harks back to Aristotle who maintains: 'For those things are natural which, by a continuous movement originated from an internal principle, arrive at some completion: the same completion is not reached from every principle; nor any chance completion, but always the tendency in each is towards the same end, if there is no impediment.'[33]

Taking up this idea that the tendency of natural forces to produce uniform results can be thwarted, and sometimes is, by hindrances of various sorts, the determinism of natural law was interpreted by Rabbi Levi ben Gershon, as only approximate and probable. He even went so far as to limit God's knowledge to the probable arguing that 'perfect knowledge of a thing is to know it as it is ... that is to know that aspect that is determined and bounded and to know also that indeterminacy which is in it.'[34]

Rabbi Levi's principle of indeterminacy did not find much favour among other Jewish thinkers as far as God's knowledge is concerned, but human knowledge, unless its source be in revelation, all agreed, is in the nature of the 'probable and plausible.'[35]

32 *Mordecai* to *Yevamot*, chap. *haḤoletz* (37b) 22. See also the commentary of Meiri to *Bava Batra* 63a, p.332

33 *Physica* II 8, 119b15

34 Rabbi Levi ben Gershon *Milḥamot haShem* III 4

35 Rabbi Ḥasdai Crescas, *Or haShem* II i, 3

6
Sampling

6.0 STATISTICAL INFERENCE

Suppose there is a given mixture of two kinds of elements, say m As and $(n - m)$ Bs making a total of n elements, so that the relative frequency of the As in the mixture is m/n. In how many ways can one draw a set of k elements from the given mixture? If one imagines a tray with k bowls in it, one for each element to be drawn, it is seen that to fill the first bowl there are n choices, since every one of the elements in the mixture is available to be picked. For each choice in filling the first one, there are $(n - 1)$ elements left to choose from in filling the second bowl, so that the first two elements can be chosen in $n(n - 1)$ ways. Similarly it can be shown that all k bowls can be filled in $n(n - 1)(n - 2) \ldots (n - k + 1)$ ways. This argument is used by the commentators on the second-century book *Sefer Yetsirah*, where substantially the same formula is given.[1]

Now consider as a reference class the collection of all $n(n - 1) \ldots (n - k + 1)$ possible ways of selecting a set of k elements. Provided only that the number of Bs in the mixture is not less than $(k - 1)$, there will be among all of these possibilities some such that the set chosen will contain just one A while the remaining $(k - 1)$

1 See below p.144

elements will be all *B*s. In fact, there are just *m* ways in which a set can be chosen having exactly one *A*, since there are *m A*s to choose from. Some possibilities will yield a set having just 2 *A*s, others 3 *A*s and, in general, just *j A*s. One can therefore compute for every *j* the number of possibilities which will yield just *j A*s. This gives then also the probability that a sample of size *k* will contain exactly *j A*s. The function thus defined is called the *probability function p(j) of the hypergeometric distribution.*[2]

For what *j* is the probability greatest? In other words, what is the most likely composition of a set of *k* elements drawn at random from the given mixture?

It can be proved that the sample whose composition is the same as the original mixture is the most likely. That is, the probability is greatest that a set will be drawn such that the relative frequency of *A*s in the sample, j/k, is as close as can be to that in the given mixture, m/n. This can be seen intuitively without computation, for whenever there is thorough mixing, such as for example in grain bins, experience indicates that a reasonably large sample usually shows almost the same proportions as the whole mixture.

Since the number of possibilities for each *j* is known, it is also possible to calculate the average value of *j*. This is called the *mean value* of the distribution and turns out to be m/n. Thus, if we suppose one sample drawn at random, the number of *A*s recorded, and the sample replaced, and then another one drawn, the number of *A*s in that recorded, and it too replaced, and so on, the average of the values of *j* recorded will tend to the mean value of the distribution.

Although the most probable composition of the sample is in the same or nearly the same proportions as the original mixture, if repeated samplings take place, the proportion of *A*s will generally vary in each sample, taking values smaller or greater than the mean, m/n.

Carnap defines:

1 A 'direct' statistical or inductive inference as one in which we

2 B.W. Lindgren and G.W. McElrath, *Introduction to Probability and Statistics*, 2nd ed. (Macmillan, 1966), p.62

conclude from the relative frequency of a given character in a population what will be the frequency of the same character in a sample drawn from that population.[3] It is then a special case of a direct inference which enables us to derive a probability from a known frequency. In that instance, the given information about a set of objects yields a conclusion about a single individual. The rabbinic principle, 'Follow the majority,' is, according to this definition, a rule of direct inference. We have already seen a more general type of direct inference used by the rabbis in the first example of Talmudic multiplication of probabilities, where the rabbis implicitly assume that the distribution in a sample of more than one element drawn from a mixture is the same as that of the original mixture.[4] However, in that context, the concept of sampling is merely incidental. In this chapter we shall explore the rabbinic treatment more extensively, and see that the Talmud also took note of the variability of the distribution.

Two further types of inductive inference are, according to Carnap,[5] the following:

2 An 'inverse' inference is one where, by virtue of the information yielded by a sample, conclusions are drawn with respect to an entire population.

3 A 'predictive' inference is one in which, on the basis of one sample, inference can be made with respect to another sample not overlapping the first.

Statistical reasoning of these types also occurs in the Talmud and instances of each will be adduced in this chapter.

6.1 DISTRIBUTION IN A SAMPLE

Our first example illustrates a direct inference and deals with *second-tithe* money. The Bible[6] commands that a second tithe be set aside from the crops and that it be taken to Jerusalem on the

3 Rudolf Carnap, *Logical Foundations of Probability*, p.207

4 See above p.51

5 *Logical Foundations of Probability*, p.207

6 Deuteronomy 14:22–27

pilgrimage festivals to serve as provisions in celebrating the holy days together with the poor and the needy. For those prevented by distance from transporting the actual grain or fruit of the second tithe, it is permitted to exchange it for coins which acquire the sanctity of the tithe and must therefore be kept separate from secular coins. This money must be taken to Jerusalem and used to purchase food for the feast. It may not be used in any other way.

The Mishnah discusses a case in which containers of second-tithe money and ordinary money broke and the coins were dispersed and mixed up, with some lost.

Ordinary money and *second-tithe* money which were scattered together – whatever is picked up singly is considered as *second-tithe* until its amount is complete, and the rest is ordinary money. If they were mixed up and could be picked up by the handful, [they are divided up] in proportion. [and any loss will therefore also be in proportion.] This is the general principle: what is collected singly goes first to *second-tithe*, but what is mixed up together is shared out in the original ratio.[7]

In the discussion on this Mishnah-text in the Jerusalem Talmud we read:

Said Rabbi Z'eira: [The rule that what is collected singly goes to *second-tithe*] is for the gain of the *second-tithe*, so that if the remainder [that has not yet been picked up] is lost, those in hand will cover it. Also, he must specify a condition, 'if those below [on the ground] are *second-tithe*, let these in my hand be in their stead.'
Said Rabbi Jonah: That is only if he picks them up singly from here and from there, but if he picked up singly all [the coins] in a [particular] area – it is as if they were mixed up and could be picked up by handfuls [and they are divided up in the original ratio].

Rabbi Z'eira makes the point that the rule which gives the benefit to the *second-tithe* is an exception based on special considerations other than the usual ones of probability. Since the *second-tithe* is sacred, it is meet that a special provision be made

7 *M. Maaser Sheni* II 5

to ensure that there will be no loss of second-tithe. However, this special provision can only be made when the coins are picked up singly, but not when they are in handfuls. This is because, when dealing with each coin separately, it can only be assigned as a whole to one or the other group. For the benefit of the *second-tithe*, it is justifiable to take the stricter view and assign the individual coin to *second-tithe*. But when they can be picked up by handfuls, such an argument cannot be justified. Rabbi Jonah adds that if some entire area is cleaned up, the coins cannot be treated as individuals even if they were picked up singly, for the distribution over the area is likely random. It is clear that both Rabbi Z'eira and Rabbi Jonah consider the ratio-rule as applicable on purely probabilistic grounds in all cases.

6.2 VARIABILITY IN SAMPLES

Because the distribution in a sample is variable, it follows that a particular sample may not exhibit the same proportions exactly as the population from which it is drawn. Indeed the larger the population, the less likely is it that the sample will have precisely the expected composition. Rather, the relative frequency in the sample will likely be near but not exactly that of the original population. Therefore, where precision is required, the sages differed as to when an inference to a sample is acceptable. The Talmud uses as a technical term the phrase 'there is homogeneity'[8] in the sense that the proportions in a sample are exactly those of the population. In cases where actual mixing occurs, as for example in liquids, it was held that 'homogeneity' can be attained. However, for solids, especially large ones, thorough mixing is harder to accomplish and furthermore samples are not likely to be large enough to justify a statistical inference, for the variability can be very great. Where to draw the boundary between what is considered small and what is large with reference to 'homogeneity,' was the subject of a difference of opinion among the rabbis. In the

8 The Hebrew phrase is יש בילה

sequel to the passage quoted in the previous section[9] these opinions are referred to.

Rabbi Yosé in the name of Rabbi Pedayah and Rabbi Jonah in the name of Rabbi Hezekiah [hold] 'there is no homogeneity except in wine or oil.' Rabbi Yoḥanan said [for objects] up till the size of olives there is homogeneity. [but not for larger ones]. Our Mishnah disagrees with Rabbi Yoḥanan – 'If they were mixed up and picked up by the handful – in proportion.' [Thus it is assumed that for coins there is homogeneity, and some coins are quite large] He can interpret it [the Mishnah, as referring to coins] up to the size of olives.[10]

The problem of variability of distribution was discussed in the earliest sources, mainly in connection with tithing. This must be done, according to scripture, 'year by year.'[11]

Rabbi Yosé ben Kippar said in the name of Rabbi Shimon Shezuri: Egyptian beans that were planted for seed, some rooted before New Year's Day – *terumah* and tithe must not be given from one lot for another, because *terumah* and tithe are not given from the new for the old nor from the old for the new. How shall he do it [since at harvest time it is difficult to separate the two]? Let him heap his harvest mixing it together, so that ... the *terumah* and the tithe will consist of new for the new and old for the old.[12]

However, this procedure was not accepted by all because it is not true that 'we can rely on mixing to say that there is the correct proportion of each in the tithe.'[13] In the Jerusalem Talmud it is suggested that only in matters of lighter consequences, for example, prohibitions that are of rabbinic but not biblical origin, may one assume that perfunctory mixing gives an accurate sample.[14]

9 See above p.81
10 *J. Maaser Sheni* II 5
11 Deuteronomy 15:20
12 *Tosefta Shvi-it* II 4 quoted in *Rosh haShanah* 13b
13 Rashi, *Rosh haShanah* 13b, s.v. ש״
14 *J. Demai* V 5

6.3 FROM SAMPLE TO POPULATION

We have used Carnap's term 'direct inference' to refer to a situation where the distribution of the population is known, and the question at hand is to ascertain the composition of the sample. The inverse problem arises when the composition of the population is not known or cannot easily be determined. Currently it is standard procedure to infer from a sample to the population and this is probably the most widely used statistical device.

In the Talmud the problem of drawing a typical example from a population of unknown composition is raised in several contexts. In general, there are two views – one maintains that two instances indicate a pattern, while according to another view, three are required, but there are some variations.

Tefillin are ritual objects which must be prepared according to exacting stipulations. One is worn on the arm and another on the head.

> If one acquires [bundles] of *tefillin* from one who is not certified, he checks two for the arm and one for the head or two for the head and one for the arm ... similarly, for the second and third bundles ... since each bundle was made by a [different] person ... the third one is mentioned to show that there is no presumption with respect to the bundles ... and even the fourth and fifth bundles [require sampling].[15]

Thus it appears that a sampling is required to establish the reliability of each individual source. However, in the Jerusalem Talmud the question is left open:

> If one found two or three bundles, he checks a pair from the first bundle and similarly from the second and third. Isaac ben Elazar asked: Is there one pattern for all of them or must it be for each one separately? If you say one pattern for all of them, he checks the first pair from the first bundle, etc., if you say each one has its own pattern, he checks three pairs in each bundle.[16]

15 *Eiruvin* 97a quoted from *Tosefta Avodah Zarah* III 2 with minor textual changes
16 *J. Eiruvin* X 1

Apparently Maimonides[17] assumes that there were bundles of standard size, for in transcribing this law, he specifies that one must take a sample from each hundred *tefillin*.

It is well known that in the Middle Ages, artisans of various kinds had certain rules of thumb for sampling. These were probably arrived at as a result of trial and error rather than by conscious experimentation and certainly not by computation of the probabilities involved. Treating the quoted instances of the Talmudic rule of sampling by threes as such a rule of thumb, it still seems worthwhile to examine its efficacy in the light of present-day statistical methods. Given a certain plan for accepting or rejecting a lot on the basis of a sample, one defines the *operating characteristic* of the plan, a function giving the probability of accepting the lot, or what amounts to the same thing, the probability that a sample will be drawn which will make the lot acceptable by our plan.[18] Thus, in our example of *tefillin* the lot size is 100, while the sample size is 3 and the plan provides for accepting the lot if there are no defectives in the sample (acceptance number $c = 0$). Now, the number of defectives in the lot may vary between 0 and 97. The operating characteristic tells us what is the probability that a lot having M defectives will be accepted on our plan of approving a lot if a sample of three has no defectives. This is determined by calculating the probability of drawing a sample of three good articles out of a lot of 100 containing M defectives. The accompanying graph shows the operating characteristic for our example. The curve falls rapidly, and the probability of accepting a lot containing 25 defectives (1/4) is only 0.42, while a lot containing 34 defectives (approx. 1/3) will be passed with a probability of only 0.28.

Now, according to the Talmudic rule cited in 3.2: 'that which is separated [from a collection] derives from the majority'[19] and in view of the fact explained there that the Talmud regards a two-thirds majority as decisive, 'by biblical law one in two is nullified,' it follows that *tefillin* would be acceptable if taken from a lot about which it is known that even up to one-third are defective, although,

17 *Mishneh Torah: Tefillin* II 10

18 See B.W. Lindgren and G.W. McElrath, *Introduction to Probability and Statistics* (2nd ed.), pp.102 ff.

19 Above p.39

of course, one would prefer the proportion of defectives to be as small as possible. Where the composition of the lot is unknown, but a sample of three turns up no defectives, the probability that the lot contains as many as 1/3 defectives is only 0.28. The probability of accepting a lot is greater than 1/2 only when the lot has less than about 1/5 defectives. The plan which prescribes an acceptance number of zero defectives for a sample of size 3 is a very effective one in light of the requirements.

Is the sampling plan prescribed by the rabbis an optimum one? To answer this question, one must investigate how the accuracy of the inference is affected by varying the size of either the lot or the sample. Obviously, increasing the sample size, for the same acceptance number zero, will reduce the probability of accepting a lot with a given number of defectives. However, if this improvement is relatively small, it will hardly be worth the trouble involved in examining the larger sample. Similarly, for a fixed size sample, increasing the lot size will lessen the accuracy of the inference but, unless this effect is relatively large for the range of lot sizes likely to be encountered, one would not be troubled too much. The figure shows the operating characteristic curves for lot sizes 10 and 100 and for sample sizes 2, 3, and 4.

The improvement in precision for increasing sample size is indicated by the shift of the curve to the left. This is important mainly in the upper part of the graph, above the line marking 0.5 probability of acceptance, since below that line the chances are that the lot will be rejected on every plan. It is clear from the graphs that, while the improvement from a sample size of 2 to that of 3 is significant, the corresponding improvement when the sample size is increased to four is relatively much smaller. Furthermore, the shapes of the corresponding curves for lot sizes 10 and 100 respectively are not much different. It appears that the rabbinic acceptance plan is quite the best for their purposes.

6.4 MORTALITY RATES

A much more interesting sampling rule is the following, dealing with the conditions that make it mandatory to declare a state of emergency on the spread of an epidemic. 'A town bringing forth

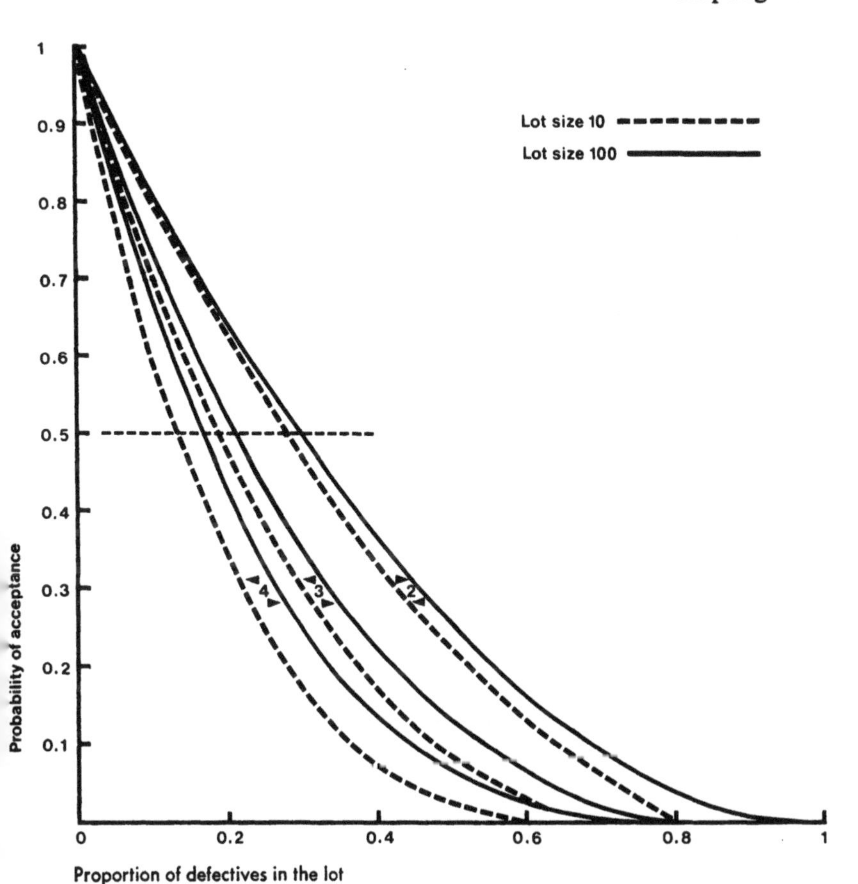

Operating characteristic curves for samples of size 2, 3, and 4

five hundred foot-soldiers like Kfar Amiqo, and three died there in three consecutive days – it is a plague ... A town bringing forth one thousand five hundred foot-soldiers like Kfar Akko, and nine died there in three consecutive days – it is a plague; in one day or in four days – it is not a plague.'[20]

Two points are worthy of note. First, it is clear that the rabbis

20 *Taanit* 21a

had some idea of a usual and normal death rate, which is proportional to the total population. A significant increase in the mortality rate sustained over a period of at least three days was cause for concern, but a co-incidence of several deaths within a short period could be expected once in a while under normal circumstances. Secondly, since there were apparently no complete census records, these being kept only for men capable of army service, the rabbis used the number of able-bodied men as an indicator of the total population. Assuming that the proportion of 'foot-soldiers' among all the inhabitants is approximately constant, the overall death rate may be computed relative to the number of foot-soldiers just as well as if the figure for the entire population were used.

The awareness that mortality follows a probability pattern and that it is the relative frequency which is the important factor was a long time in developing elsewhere. By contrast, it is interesting that David remarks in discussing the work of the seventeenth-century demographers in London: 'The years were classified according to the number of plague deaths; if the number of these was greater than two hundred in any particular year, and the yearly deaths exceeded those of the preceding years, that year was called "sickly." Since the population of London was increasing rapidly all the time it follows that an increasing number of years would be sickly.'[21]

6.5 PREDICTIVE INFERENCE

Carnap distinguishes between an 'inverse inference' from a sample to a population and a 'predictive inference' from one sample to another sample not overlapping the first.[22] The latter really consists of an inverse inference followed by a direct one. As such, of course, it involves greater uncertainty.

A far more important source of difficulty is the fact that one generally resorts to predictive inference in situations that differ in

21 F.N. David, *Games, Gods and Gambling* (New York, 1962), pp.105–6
22 Rudolf Carnap, *Logical Foundations of Probability* (2nd ed.), p.207

a fundamental way from those in which sampling of small populations occurs as illustrated above. Thus, even in the simple rabbinic examples to be discussed below, it is not a fixed population that is dealt with. Rather one is concerned with an open-ended series of events, and on the strength of past experience, one seeks to forecast the future. If a certain treatment has been found to be effective in some cases of a disease, may one confidently expect that it will be universally effective, or even the very next time it is tried? How many past successes justify anticipating future success? Is there any reason to believe that our experiments yield a random sample of the kind of events they are? When one samples the objects in a particular lot, a precise definition of randomness may indeed be hard to come by, but somehow one feels that any object is as likely to be drawn as any other, and so a probability interpretation either in terms of empirical frequencies or logical alternatives seems acceptable. In the case of temporal events, such a conviction is not easily attained.

The heart of the matter is the traditional problem of induction, which is just the search for some principle that justifies extrapolating from past observations to the future and specifies acceptable ways of doing it. Nonetheless, its practical importance is such that we find 'predictive inference' already in the Talmud. We have cited a case of prediction dealing with the probability that a woman will bear a live male child.[23] Our next example deals with medical practice and emphasizes the acute need for inductive reasoning in coping with the most urgent demands of life and health.

The physicians of antiquity often prescribed the wearing of amulets containing herbs or writings of various kinds. Since the function of such amulets was not understood, the rabbis were fearful lest they be works of magic rather than medicine. On the other hand, they regarded the practice of healing as a sacred obligation. Therefore, they devoted considerable attention to defining the border line between science and sorcery. This was concisely formulated as follows: 'Abaye and Rava both said: Whatever has healing effect is not included in the "ways of the Amorite" [i.e., sor-

23 Above p.50

cery, which is forbidden]. That which does not have healing effect is included in the [forbidden] "ways of the Amorite." '[24]

It thus became necessary to define criteria for evaluating particular prescriptions as well as for authorizing practitioners. Thus, although approved amulets may be worn even on the Sabbath, the Mishnah forbids this for 'an amulet that is not from an expert.'[25] But what constitutes a standard for approval?

The rabbis taught: What is an approved amulet? – One [prepared according to a formula] that has healed [once], a second time and a third time; whether it is an amulet in writing or an amulet of roots, whether it is for an invalid whose life is endangered or for an invalid whose life is not endangered ...

But it was taught: What is an approved amulet? – One that has healed three people simultaneously. [Is there not a contradiction between these two rules?]

There is no difficulty: The one [rule] is to authorize the practitioner [i.e. one that has simultaneously healed three people suffering from different ailments by prescribing three different amulets, is recognized as an expert and his amulets are acceptable]; the other [rule] is to approve the amulet [i.e., a formula which has cured the same complaint three different times is approved for general use].[26]

However, this question continued to trouble the rabbis for many generations, as can be seen from the following passage in a letter by Rabbi Yedaiah ben Abraham Bedersi (1270–1340):

Our great master [Maimonides, Guide to the Perplexed III, 37] has explained ... that whatsoever follows logically from scientific premises may be done for any beneficial purpose ... and certainly for human healing ... However for medical benefits for the human body, even if it is not implied by reason, we permit it if its usefulness is well-known at the time, such as the function of ... remedies which have been tested many times and their usefulness is well-known ... It follows then from his words that

24 *Shabbat* 67a, See also *Ḥullin* 77b
25 *Shabbat* VI 2
26 *Shabbat* 61a, b

that which is neither implied by reason nor are its benefits well-known, is clearly prohibited and is accounted as sorcery ... How were all of today's well- known remedies discovered if not by experiment? And, shall we now close the gates of experimentation? ... Furthermore, it is necessary to examine what is considered 'well-known' in matters which are not implied by reason – does it mean that most books written on the subject mention it and unless most authors agree on it, it is not well-known; or is it 'well-known' if one author alone attests to it provided that he is famous in scholarship, or is it 'well-known' even if it is not a sage who testifies to it, or even just some old women who have many vain traditions concerning remedies which are not scientific nor known to the experts – all this we need to examine ... [27]

The question 'what proportion of successes justifies a predictive inference?' is still with us.

6.6 STATISTICAL LAWS OF NATURE

We have seen that even in the classical interpretation of probability which is applicable to individual objects or events, the crucial notion is that of a collective, namely the set of all possible outcomes. It is a characteristic of probability-type thought that it concerns itself with collections of elements rather than with individuals. It is to be expected, therefore, that one should find in the rabbinic sources applications of statistical thinking to conceptions of the laws of nature and society.

Aristotle viewed nature as the function of 'an internal principle' in every individual object. Consequently, he maintained, 'In natural products the sequence is invariable, if there is no impediment.' In a collective, if some individuals fail to conform to the general pattern of 'nature,' this must be attributed to some impediment which results in a 'mistake,' for 'clearly mistakes are possible in the operations of nature also.'[28]

On the other hand, Maimonides suggests that there may be reg-

27 The letter was written in reply to a proclamation of Rabbi Shlomo ben Adret in 1305, and is included among his *1255 Responsa*, no.418

28 *Physica* II 8, 199a35 ff.

ularities that apply to entire groups as a whole but which do not necessarily arise from 'an internal principle' in every single member of the group. In this manner, he attempts to resolve the apparent contradiction between certain prophetic statements and the principle of free will. He interprets the prophecy as a statistical forecast, which still permits the individual his freedom.[29]

Is it not written in the Torah: 'They shall serve them, and they shall afflict them'?[30] Was this not a decree that the Egyptians should do evil [and enslave Israel]? It is also written 'This people will rise up and go astray after the foreign gods of the land.'[31] Was this not a decree that Israel should worship idols? Why then did He punish them? The reason is that He did not decree concerning any particular individual that he go astray. Any one of those who went astray and worshipped idols, had he not desired to commit idolatry would not have done so. The Creator only informed him [Moses] what is the normal course of the world, as one might say 'This people will have among them righteous men and evildoers.' No sinner can claim on this account that it has been decreed that he be wicked since the Holy One Blessed be He informed Moses that there would be wicked men in Israel. Similarly, it is written 'For the poor shall never cease out of the land'[32] [but this does not imply that any particular individual is pre-ordained to be poor]. So too with the Egyptians. Everyone of those who oppressed and ill-treated Israel had he not wanted to harm them was free to refrain, for He did not decree concerning any specific person.

Later, Rabbi Isaac Aramah applied the same idea to biological law. He considers the instinct for self-preservation to be part of the nature of all living things, yet he sees it as being a statistical law which is operative only in sufficiently large assemblages of individuals.[33]

29 *Mishneh Torah: Teshuvah* VI 5
30 Genesis 15:13
31 Deuteronomy 31:16
32 Ibid., 15:11
33 *Akedat Yitzḥak*, chap.99

The law of nature implies that no being will sever itself from its life. It is a covenant with every living thing that it will not destroy itself.
Though it is conceivable that 'one from a city'[34] in the multitude of days because of a deteriorated imagination will transgress that law and destroy himself, nonetheless, this possibility cannot be conjectured for all the individuals of a species, as [the Talmud] says: 'his attitude can be disregarded among all men' ...[35] the more so will the nature of this matter be strengthened and firmly established forever if the individuals in the covenant are very great in number ...
... It is as if it was conjectured that one day all men should choose to strangle themselves. Now that which is not in the nature of man to do is considered an absolute natural impossibility. Similarly it is a covenant with the world that its population will not cease by choosing to desist from procreation or in some similar manner, although it is possible that 'one from a city' will so choose.

It is worth noticing that Aramah exphasizes that a statistical law is better established if the class to which it applies is 'very great in number.' This is really a consequence which flows directly from the Law of Large Numbers which Aramah himself formulated.[36]

34 The phrase is from Jeremiah 3:14 and is intended to convey the sense of a small minority.
35 *Berakhot* 35b
36 See above p.27

7
Paradoxes

7.0 LOGICAL DIFFICULTIES

Although a definition of randomness is never, to my knowledge, attempted by the rabbis, there is a very enlightening discussion in the Jerusalem Talmud, where it is made clear that mere ignorance or lack of evidence does not warrant assigning equal probabilities to logical alternatives, although that is the practice in purely 'chance' or 'random' situations. In this text two paradoxes are propounded – one to show that a situation, where logical alternatives exist but nothing is known about their distribution, cannot be treated as equivalent to one where a random distribution can be assumed; and another, to demonstrate that acceptance rules based on observed or known frequencies can lead to contradictions when applied to individuals.

Our text deals with partnership when the shares of the partners are unequal.

We learnt: Three put [coins] into a money-bag, and part of it was stolen – [the remainder is divided between them proportionately]. Is that how they divide it? Have we not learnt concerning these stones [of a collapsed two-storey house and each storey belonged to another] that if [some of the building stones] were stolen – half is of the one and half of the other [regardless of the proportions of ownership]? Said R. Shammai, building stones are large [and when one takes them, it is one at a time] and it is not known whether he took from this one's or from that one's, and because of the doubt – half is of the one and half of the other. But coins are small and it is possible to mix them up evenly – to do justice to all, each one takes according to his investment. On what grounds do you say that we consider the stolen ones [and each owner suffers half the loss in the case of building stones]; perhaps we should consider the remainder?

Said Rabbi Yosé ben Rabbi Bon: Even if we consider the stolen ones, justice will suffer [under the rule that the status of each individual stone is decided separately]. Suppose the one has fifty and the other a hundred,[1] and fifty were stolen – the owner of the fifty will not lose anything [because when each stone, down to the fiftieth one is picked up by the thief, the likelihood is greater that it belongs to the owner of the hundred].[2]

It is here clearly recognized that where thorough mixing occurs, as in the case of three partners A, B, C putting n_A, n_B, n_C coins respectively into a bag, it is reasonable to assume that a random selection of m coins will reflect the composition of the original group. Now this is based on the assumption that the drawing of any particular coin is equally likely, regardless of whether it belongs to one or another of the partners. Ownership and even physical differences between the coins are all regarded as irrelevant insofar as the probability of drawing them at random is concerned. Therefore, this case is treated as if nothing were known which is likely to affect this probability. It follows then that the probability p_A of a given coin belonging to A is $n_A/(n_A + n_B + n_C)$ etc., i.e., $p_A : p_B : p_C = n_A : n_B : n_C$.

However, in the case of large building stones, which cannot easily be distributed evenly, other contingencies must be taken into account.

1 Some editions read 'a hundred and fifty,' instead of 'a hundred'

2 *J. Bava Kama* IV 1. It is usual in the Talmudic literature to cite sources merely by the opening phrase and the reader is expected to complete the quotation. Quotations from the works of the *Tannaim*, are indicated by the introductory formula 'We learnt.' Thus our passage opens with the formula 'We learnt' followed by the phrase 'Three put into a money-bag, and part of it was stolen.' The source of this citation is *Tosefta Ketubot* x 5, and it has been completed accordingly.

The interpretation of this entire passage follows that of the commentary by Rabbi Jacob David Ridbaz (1845–1913).

It may very well be that all or most of the exposed and easily accessible stones on the heap come from the upper part of the wall, if it collapsed under its own weight, or the reverse might be true in other circumstances. It may be that the thief has reason to prefer stones of a particular size. Other possibilities, too, suggest themselves. Consequently, it cannot be maintained that the probability of drawing any individual stone is the same. In fact, we know nothing at all about the probability p_A of drawing a stone belonging to A as compared to the probability p_B of drawing a stone belonging to B. Rabbi Shammai suggests therefore that, in view of our ignorance, we may assume that $p_A = p_B = 1/2$.

In modern times, Rabbi Shammai's proposal appears as the 'Principle of Indifference.' By virtue of this principle, it was argued that 'each proposition of whose correctness we know nothing is endowed with a probability of 1/2, for the proposition and its contradictory proposition can be regarded as two equally probable cases.'[3]

7.1 A PARADOX OF INDIFFERENCE

To counter Rabbi Shammai's 'Principle of Indifference,' the Talmud shows that it leads to a paradox. For if one considers the stolen stones and assumes that the probability is equal that a particular stone belongs either to A or B, the total loss is suffered equally by both partners despite the fact that their shares were unequal originally. For definiteness, let the shares of A and B be 100 and 50 respectively and suppose 50 were stolen. Each one will therefore lose 25 and their portions will now be 75 and 25 respectively.

On the other hand, one ought to be able to apply Rabbi Shammai's reasoning, if it is valid, to any stone, and therefore also to the remaining stones. Thus, for any one of the remaining stones the probability should be equal that it belongs to either A or B. In

3 Richard von Mises, *Probability, Statistics and Truth*, p.11, paraphrasing J.M. Keynes, *A Treatise on Probability*, p.42. See above p.63

that case, A and B would each get 50, which contradicts the first calculation. Moreover, if one considers the remainder, division into half and half cannot be applied for another reason as well. For, of the 100 stones remaining, at least 50 certainly belong to A, since B never had more than 50 altogether. A's share can vary then between 50 and 100. How then can one say that the chances are equal for each stone that it belongs to A or B?

One could, of course, pick 50 stones from the 100 and assign them to A on the basis that any one of the stones is equally likely to be one of the 50 certainly belonging to A. Then the remaining 50 could be divided equally between A and B, and this would accord with the first procedure. This amounts to dividing up the original 150 stones into three groups and treating each separately. Thus, for the stolen 50, the probability that a given stone belongs to A is assumed to equal the probability that it belongs to B. As for the remaining 100, since 50 of them are known for certain to belong to A, one is forced to eliminate 50 from the dispute. The 50 that are left are then treated like the stolen ones.

To this it can be objected that one might just as well treat all the stones alike initially, arguing as follows. There certainly were 100 that belonged to A. Since there is no reason for favouring any particular stones above others, all 150, including the stolen ones, are equally likely to belong to this group of 100. Thus 2/3 of the stolen ones will be assigned to A, and so the losses will be proportional to the original shares.

On this scheme too, a principle of indifference is used, and what reason is there for preferring the first procedure over this one or vice versa?

7.2 AN ACCEPTANCE PARADOX

Rabbi Yosé ben Rabbi Bon poses another puzzle. He is prepared to admit that proportional allocation of the loss is not justified. This is because it is not at all likely that the sample which the thief got is representative of the ownership distribution in the original pile of stones. Nonetheless, the situation is not one in which there is no evidence at all to go on. Nothing is known, it is true, about how the thief picked his stones, but regardless of what criterion applies, it is not possible that no effect at all can be ascribed to the known preponderance of stones belonging to one owner. Since one cannot use that principle of statistical inference which yields a conclusion for the entire set of stolen stones, why not use a simpler acceptance rule which gives an inference for a single instance? It is legitimate to argue that the probability is greater that a particular stone taken from the heap belongs to the majority. One cannot assign the value 2/3 to this probability just because the owners' shares are in a ratio of 2:1, nor may one assume with Rabbi Shammai that it is exactly 1/2. Now, on an ordinary acceptance rule, this information is sufficient to warrant assigning the stone in question to the majority partner, since in fact the stone as a whole must belong to one or the other. However, this too must lead to a paradox, for the same reasoning can be applied to each one of the 50 stolen stones, and in every case there is a probability greater than 1/2 leading to the same conclusion, namely that that stone belongs to the majority owner. The conjunction of these inferences, that all the stolen stones came from the one partner's share only is highly implausible and 'justice will suffer.'

There is here the clear recognition that the probability $p_A < 1$ assigned to a single event is really a measure of the relative frequency of the given event recurring in repeated trials, whenever such a sequence can exist. With respect to an individual, the rule 'Follow the Majority' is an extrapolation which cannot be applied indiscriminately. In a comment of the Tosafot in another context, the argument is very bluntly put: 'We may not act falsely because of the majority, for there certainly is one not [of the majority].'[4] Each individual inference by the majority acceptance rule is subject to error and the conjunction of all of them is certainly false.

We have already met with an example in the case of mixtures, where this idea played an important role.[5] In general, as we have seen, where a number of elements of the mixture equal to the forbidden ones have been destroyed, there is less uneasiness, since the certainty of error is gone, for it is not a contradiction to assume

4 *Zevaḥim* 71a, end of s.v. אפילו
5 See above p.41

that all the prohibited ones are gone. Nonetheless, even then, on probability grounds, a problem remains, for the likelihood is small that the lost ones were all the forbidden ones.

An interesting solution is given in the Talmud which is applicable in some types of cases. Our example deals with the rite of the burnt-offerings in the Temple. Any blemish or imperfection disqualifies an animal for a sacrifice and it is therefore prohibited to place it on the altar. The Mishnah speaks of the parts of one blemished animal having been mixed up with many similar parts of fit animals that were waiting to be put onto the altar. 'Rabbi Eliezer says: If one head has already been offered [before the facts were known, so that we are now concerned only with the incomplete mixture], then all the heads may be offered; if one pair of legs have already been offered, then all the legs may be offered.' In the Gemara we read: 'Rabbi Elazar said: Rabbi Eliezer permitted it only two [heads or pairs of legs] at a time, but not singly.' For, explains Rashi, 'one of the pair is certainly a permitted one, so that the act of putting onto the altar is allowable for the permitted one, but as for the forbidden one – one can suppose it is the one that was already offered.'[6]

In other words, by virtue of the fact that one of the pair is certainly acceptable, the act of presentation is warranted. As for the possible objection that one of the pair may be the forbidden one, that is effectively neutralized by the argument that the likelihood of any particular one being from the blemished animal is no greater than the likelihood that the one already offered was from that disqualified animal. Thus there is no decisive reason to ban the act in question, whereas the reason for requiring it is certain.

7.3 BERTRAND'S PARADOX

These ancient paradoxes are the bane of modern conceptions of probability as well.[7] The counterpart of the first is generally known as Bertrand's Box Paradox. Although it is not quite the same as the

6 *Zevaḥim* 77b

7 Keynes has collected a number of paradoxes generated by the Principle of Indifference in chapter IV of *A Treatise on Probability,* pp.41–51

Talmudic one, there are elements of similarity. The second, unchanged in essence, appears in modern garb as the 'lottery paradox' and is still the subject of heated debate.

The following is a common version of Bertrand's paradox.[8] Consider a box with three drawers. In one drawer there are two gold coins; in another, two silver coins; and in the third, there is one silver and one gold coin. A drawer is chosen at random and a coin is drawn at random from that drawer. The selected coin is gold. What is the probability that the other coin in the same drawer is gold?

There are two contradictory answers. One might argue that since the selected coin is gold, the drawer it came from cannot be the one containing two silver coins. It must be either the drawer with the two gold coins or the one with a gold and a silver coin. There are then two possibilities with no reason to prefer one over the other, and so the Principle of Indifference applies. On one alternative the remaining coin is gold while on the other it is silver. Therefore the sought-for probability is 1/2.

Another answer requires that all possible alternatives be treated equally to begin with. There are, in fact, six possibilities:

1 the drawer with two gold coins is chosen and the first coin is drawn
2 the same drawer is selected and the second coin is drawn
3 the drawer with two silver coins is chosen and the first coin is drawn
4 the same drawer is chosen and the second silver coin is drawn
5 the drawer with the mixed coins is selected and the gold coin is drawn
6 the 'mixed' drawer is chosen and the silver coin is drawn

Of these six alternatives, three are ruled out by the knowledge that in fact a golden coin was drawn. Thus cases 3, 4, and 6 did not occur, since they could not yield the coin that was actually drawn. There remain, then, three possibilities of which two are favourable to the remaining coin being gold. Consequently, the probability is 2/3 that the second coin is gold.

8 See Henry E. Kyburg, Jr. *Probability an ıductive Logic* (Macmillan, 1970), p.34

The use of the Principle of Indifference seems justified in both answers although they are obviously inconsistent. The difference between the two methods of analysing the problem lies in the fact that the second treats all the ultimate possibilities alike, whereas the first proceeds to answer the question: 'Which drawer is the source of the gold coin that has actually been drawn?'

A similar remark holds for the Talmudic paradox. If one asks 'Where does this stone come from?' and applies the Principle of Indifference, in the end one must eliminate a whole group of stones from consideration. On the other hand, if every possible way of drawing the stolen stones is regarded as equally likely, one ends up with a sample distribution like that in the original population.

7.4 THE LOTTERY PARADOX

Rabbi Yosé's paradox of the Acceptance Rule in its latter-day form is of very recent vintage and appears to have been originated by Henry E. Kyburg.[9] The question arises naturally in a logic of induction: is there ever ample justification for *accepting* an hypothesis, even tentatively? If yes, then there must be an acceptance rule which sets forth the conditions under which a given proposition may be taken as true. That most people do, in fact, act on assorted propositions with complete confidence is a truism. Thus most 'observation statements' which report sense experience are regarded as practically certain, as are also various laws of physics and chemistry.

Now, consider a fair lottery in which the prize is to be awarded to one winner out of a million ticket-holders. The probability that ticket number 3, say, will win is only one in a million, and the probability that it will not win is overwhelmingly large, namely, 0.999999. If there is any acceptance rule which commends itself, certainly such a high probability should be enough to warrant

9 See Henry E. Kyburg *Probability and the Logic of Rational Belief* (Wesleyan U.P., 1961), p.197. According to Keith Lehrer, this is the first appearance of the paradox in the literature ('Induction, Reason and Consistency,' *British Journal for Philosophy of Science* 21 (1970), 103.) See also, Isaac Levi, *Gambling with Truth* (New York: Knopf, 1967), pp.38 ff.

accepting the conclusion that ticket number 3 will not win. The same argument, though, applies to every single lottery ticket, so that for *i*, any number between 1 and 1,000,000 inclusive, it follows that ticket number *i* will not win. If all of these conclusions are accepted as true, one cannot fail to accept also their conjunction, that no ticket will win, which is obviously false, since in a fair and honest lottery one ticket does win.

How to frame acceptance rules that will avoid the paradox is a challenge that is exercising many logicians,[10] although some would rather do without the rules themselves.

7.5 WHENCE PARADOX?

We return to the Talmudic 'acceptance paradox.' The source of the difficulty is that it is necessary to operate at two distinct levels. Insofar as the dichotomy permitted/forbidden is concerned, there are only two mutually exclusive alternatives which are unequally weighted: it is therefore reasonable to suppose that the more probable alternative is to be accepted. Thus suppose there are m permitted objects and $(n - m)$ forbidden ones. The set P consisting of the m permitted objects is thought of as an alternative of weight m/n, and the other alternative F has weight $(n - m)/n$. On the level of individual drawings, however, we have equipossible alternatives which number as many as the elements in the mixture. To the question, 'Which particular individual will be drawn on a given single draw?' there can be no sensible answer, for any one is as likely to be drawn as any other. Moreover, when one is actually drawn, it is not warranted to conclude that this particular one was more likely to be drawn than any other one.

This is a crucial point on the view that identifies equipossibility with equiprobability; and, on the level of individual drawings from

10 See, for example, J. Hintikka and R. Hilpinen 'Knowledge, acceptance and inductive logic,' in *Aspects of Inductive Logic*, edited by Hintikka and Suppes (Amsterdam, 1966), pp.1–20. Also, Risto Hilpinen *Rules of Acceptance and Inductive Logic* (Amsterdam, 1968). An elementary survey of various approaches is found in H.E. Kyburg, *Probability and Inductive Logic*, chap.14, pp.180–98.

a collection, the frequency interpretation also accepts as basic the fact that every individual has an equal chance to be drawn. Drawing a black object, say, from an unknown mixture, increases the probability that may be ascribed to the set of black objects, but it in no way affects the probability of any one particular object.

This idea is poignantly emphasized by Maimonides in another connection: 'This [is] the nature of the possible. For it is certain that one of the possibilities will come to pass. And no question should be put why one particular possibility and not another comes to pass, for a similar question would become necessary if another possibility instead of this particular one had come to pass. Know this notion and grasp it.'[11]

Since every object drawn is an element in one or the other of the sets P and F, there is a predetermined many-to-one correspondence in which every individual drawing is assigned to one of the alternatives P or F. The rule of inference, 'Follow the Majority' is essentially an attempt to approximate this existing partition. Inasmuch as the individual objects are indistinguishable, the assignment of any particular drawing alone to either one of P or F cannot lead to contradiction. On the other hand, neither more nor less than m drawings can be assigned to P. To do so is inconsistent with the premises. Now supppose the n objects to be labelled x_i $(i = 1, 2, \ldots, n)$; it is justifiable to say $x_i \in P$ (read: x_i belongs to P) for any single i. However – and this is what distinguishes an inductive inference from a deductive one – having chosen a particular value of i, say i_0, such that one asserts $x_{i_0} \in P$, there are only $(m - 1)$ choices left from among the remaining $(n - 1)$ values of i for which $x_{i_0} \in P$ can be asserted. Once all these these choices have been exhausted there remain $(n - m)$ drawings which cannot be assigned to P.

On the other hand, the difficulty can be overcome by making assertions, not about individual drawings, but only about sets of no less than $(n - m + 1)$ drawings. Every such set, it is true, contains members of P.[12] This is, of course, the solution proposed by

11 Moses Maimonides, *Guide of the Perplexed* III 26 (Pines, p.509)

12 Cf. a *Confidence Interval*, which is a selected range of possible values for a parameter that is asserted to contain that parameter. See B.W. Lindgren and G.W. McElrath *Introduction to Probability and Statistics*, pp.174 ff.

Rabbi Elazar who suggests offering two similar parts of the sacrifices simultaneously. In fact, this is really a deductive conclusion and it is therefore universally valid. However, like all deductive inference, its weakness is that it tells no more than is already contained in the premises. One turns to inductive reasoning precisely because nothing is known about the individual drawings (or the sets of less than $n - m + 1$) and it is conclusions about those that are needed. To preserve consistency, one is then perforce precluded from drawing universally valid conclusions and is limited to only m individual assertions, of which moreover no single one is certain. Nonetheless, they are all free from the danger of contradiction.

Now, as we have seen, the Talmud chooses this approach in the case of mixtures, where once a number equal to the forbidden ones is no longer extant, one can safely 'follow the majority.' The same solution was used in another instance. Among biblical incidents described in the Talmud as involving the use of lots is the exchange of the first-born Israelites for the Levites. The relevant passages are quoted above in chapter 2, Random Mechanisms.[13]

There were 22,273 ($=N$) first-born sons, but only 22,000 could be redeemed by exchange for an equal number of Levites. The remaining 273 would have to pay five shekels each. Opponents of a lottery might argue that choosing those who should pay by chance is not fair, since the debt ought to be regarded as one to be paid by the entire group rather than by individuals selected by an arbitrary method. However, if some method could be devised whereby the Divine will might assert itself and unambiguously designate those who are to pay, all objections would be silenced. The Talmud suggests that the number of lots marked 'Levite' was equal to the number of first-born sons (N), and the additional 273 lots were marked 'five shekels.' From this total of $N + 273$, only N were drawn. Thus, those who drew a ballot marked 'five shekels' could not argue that the lot could not consistently express the Divine will, if such it were, to release all the first-born from the

13 See pp.29 ff.

obligation to pay. It was logically possible to draw N lots that did not require payment of five shekels. However, freedom from contradiction is only a minimal requirement. A set of statistical inferences that are not inconsistent with the premises and with each other is not certainly false; but neither is it necessarily true. Where the consequences of error are great – and injustice undoubtedly falls into this category – nobody ought to be penalized on the grounds of a merely consistent inference. To conclude that these particular people are to pay five shekels, evidence is required that their drawing lots so marked is actually an indication of God's will. If, however, 'they came up at regular intervals,' that is so improbable an occurrence that the hypothesis of Divine intervention cannot be dismissed.

What is actually accomplished by increasing the number of lots beyond those which are to be drawn? If we apply this idea to the Kyburg lottery paradox, it means that there would be a million and one tickets, one of them being a blank, so that if it were drawn, no prize would be awarded. This contradicts the premise that the lottery is 'fair,' namely, that one (numbered) ticket will certainly win. Yet, if we assume a quantitative measure of 'fairness' in terms of the probability of the proposition that 'one of the numbered tickets will win,' it is clear that under the new scheme, the decrease in 'fairness' is negligible indeed. Instead of a probability of 1, we now have 1 reduced by one part in a million and one. The probability of the conjunction of all the statements, 'ticket number i will not win' (for $1 \leqslant i \leqslant 1{,}000{,}000$) is, however, no longer zero, although it is very close to it. It is, therefore, not a demonstrably false conclusion that is given by the conjunction of the individual statistical inferences. Yet, as we have seen, the rabbis were not satisfied with the mere elimination of contradiction. The essence of the paradox is really that, granting the validity of statistical acceptance rules, the conjunction of statistical inferences should be itself acceptable by the standards of the assumed acceptance rules. Thus it is not sufficient that the conclusion 'no numbered ticket will win' is possible. The paradox is rooted in the fact that 'no numbered ticket will win' – a conjunction of statements of high probability – itself has a very low probability.

7.6 SHRINKING PROBABILITIES

This problem is explicitly formulated by the medieval Talmud expositors. It arises in considering a majority of a majority. For example, the Talmud[14] argues that the majority of women who conceive give birth to a viable child, while the collective of women who conceive is itself a majority of all women. Since we are dealing here with uncounted majorities, how are we to decide whether the majority of the majority is not actually a minority of all women? Rabbi Yom Tov ben Abraham puts it thus: 'Perhaps you are troubled that in this entire discussion [the Talmud] refers only to the minority that abort, but why do we not take into account also the minority that do not conceive at all, thus increasing the measure to the point that the possibility of a live birth can be overlooked [being in the minority]. The answer is that the minority that do not conceive is a small minority and we count it together with the minority that abort to make out of both together a substantial minority.'[15]

This explanation may suffice for the particular text but, with repeated multiplications, the issue must ultimately be faced of ascertaining when the majority ceases to be a majority.

Another radical solution refers to a Talmudic statement that under certain circumstances 'the minority is [considered] as if it were not,'[16] and it is suggested that perhaps in such cases, once having accepted a proposition as valid because of a sufficiently large majority, one then operates with it as with a certainty.[17] However, neither of these explanations seems convincing to Rabbi Shlomo ben Adret, who concludes that the matter 'requires study.'[18]

On the other hand, in dealing with measured mixtures the Mishnah rules: 'a bushel of *terumah* fell into a hundred: he lifted

14 *Yevamot* 119a

15 Rabbi Yom Tov ben Abraham, Commentary to *Yevamot* 119a, s.v. אלא

16 *Niddah* 18b: 'In three instances the sages followed the majority treating it like a certainty.'

17 Rabbi Mosheh ben Naḥman in his commentary to *Ḥullin* 86a, s.v. הא

18 Commentary to *Yevamot* 119a, s.v. אלא

out [one of the mixture], and another one [of *terumah*] fell in, and so on – it is permitted until the *terumah* is more than the secular.'[19] Since the relative frequencies are known, it is possible to compute precisely when the proportion of *terumah* in the mixture becomes a majority. It seems that, in the main, the rabbis acknowledged the limitations inherent in statistical acceptance rules, and resigned themselves to them.

Rudolf Carnap defends a similar position. For the determination of a rational decision, the acceptance of some new proposition is not necessary. If the evidence makes the proposition credible to a sufficient degree, one may then reasonably act upon it.[20]

19 *M. Terumot* v 7
20 Rudolf Carnap, 'The Aim of Inductive Logic,' Proc. 1960 International Congress – Logic Methodology and Philosophy of Science, p.317

8

Evidence and entailment

8.0 TWO TYPES OF PROBABILITY

Rudolf Carnap[1] maintains that there are two basically distinct concepts covered by the word 'probability.' In order to avoid semantic confusion, he refrains from introducing special terms for these different concepts, and prefers rather to designate them *probability*$_1$ and *probability*$_2$. The latter is identified with relative frequency, whereas the former is much broader. To explain it some preliminary remarks are required.

When propositions are deductive consequences of known premises, the logical relation between the premises and the conclusion is one of implication. Given the premises, the conclusion cannot be otherwise. However, most hypotheses cannot be deduced from premises known to be true. Consequently, the hypotheses themselves cannot be established beyond doubt. Nonetheless, hypotheses can be confirmed by evidence of various sorts. There is then some kind of logical relationship between evidence-statements and conclusion-statements by virtue of which the evidence endows the conclusions with partial validity. The relation between them is one of 'confirmation.' Whereas in deductive logic the premises imply the conclusion, in inductive logic the premises confirm the conclusion.

1 Rudolf Carnap, *Logical Foundations of Probability* (2nd ed.), chap. II, pp. 19 ff.

The concept of confirmation may be *comparative* in the sense that one can speak of a certain hypothesis h being more strongly confirmed by the evidences (e_1 & e_2) than by e_1 alone. A *quantitative* or *metrical* concept of confirmation enables one to assign a measure and say that e_1 confirms h to the degree r, where r is some number between 0 and 1. Probability$_1$ is identical with the concept of confirmation, and may be conceived of on three levels:
(i) *Classificatory*: 'the hypothesis h is confirmed by the evidence e' equals 'e renders h probable.'
(ii) *Comparative*: (a) 'h_1 is confirmed by e_1 more strongly than h_2 by e_2' equals 'e_1 renders h_1 more probable than e_2 renders h_2.' (b) 'h is confirmed more strongly by (e_1 & e_2) than by e_1 alone' equals '(e_1 & e_2) render h more probable than e_1 alone does.' (c) 'e confirms h_1 more strongly than it does h_2' equals 'e renders h_1 more probable than it does h_2.'
(iii) *Quantitative*: 'h is confirmed by e to the degree r' equals 'the probability of h on the evidence e is r.'[2]

We have already seen that the rabbis drew conclusions from statistical data on which to base decisions where relevant sequences or classes yield information about probabilities. However, these are not the only sources of probability assignments. The concept of legal evidence is much broader, and anything that lends credibility to a statement endows it with probability.

From frequencies or from an enumeration of equiprobable alternatives, quantitative probabilities can be derived and, as shown above, the rabbis even devised a calculus of probabilities. For evidential statements of a general nature, there is no unanimity among philosophers as to whether a metrical concept of probability is at all appropriate, although in modern times the suitability of a comparative concept is not questioned.[3] In this chapter we shall investigate the rabbinic approach to evidence and how it can be measured, and how probabilities based on a very comprehensive class of evidence can be compared with that computed from relative frequencies.

2 Rudolf Carnap, *Logical Foundations of Probability*, p.163. See also preface to the second edition p.xvi

3 Ibid. This was not always so. See below p.156

8.1 A COMPARATIVE CONCEPT OF EVIDENCE

The crucial distinction between a comparative concept of evidence and one that is merely classificatory comes to the fore in cases where for a given proposition there is an accumulation of evidence, each datum of which is independent of and not deducible from the others. Does every additional item of evidence increase the probability of the hypothesis under consideration? If yes, the scheme is comparative.[4]

An illustration of a sequence of evidential statements is the following incident recorded in the Talmud[5] about the president of the Sanhedrin early in the first century B.C.E.

> Rabbi Shimon ben Shetaḥ said: May I see the consolation as I saw a man pursuing his fellow into a ruin, and I ran after him, and I saw a sword in his hand dripping blood and the slain man writhing [in convulsions]. I said to him: O wicked man, who killed this one, either I or you! But what shall I do since your blood is not given into my hand, for the Torah said: 'On the evidence of two witnesses ... shall he who is to die be put to death' ...[6]

We have here a series of observations, each of which increases the probability of the conclusion, and when all of them are added together with the final one that nobody else was present, the identity of the murderer appears incontrovertible. Yet Rabbi Shimon ben Shetaḥ could not testify that he saw the murder committed. In fact, the Talmud says that this story was used to explain to witnesses what kind of observation is not trustworthy in capital cases. 'If you will testify on probability ... know that his blood and the blood of his progeny will be on [you] forever,' the witnesses were warned, and 'probability' was defined for them by citing this kind of occurrence reported by Rabbi Shimon ben Shetaḥ.[7]

4 See Carl G. Hempel 'Fundamentals of concept formation in empirical science,' *International Encyclopedia of Unified Science* (Chicago, 1952) vol.II, no.7, 77

5 *Sanhedrin* 37b, quoted from *Tosefta Sanhedrin* VIII 2

6 Deuteronomy 17:6

7 *Sanhedrin* 37a

Maimonides refers to this example in his comments on the inadmissibility in criminal cases of even the most convincing circumstantial evidence.[8] In the course of his remarks he points out that there are gradations of probability, corresponding to the cumulative effect of the sequence of evidences.

The 290th commandment is the prohibition to carry out punishment on a high probability, even close to certainty ... Do not think this law unjust. For among contingent things some are very likely, other possibilities are very remote, and yet others are intermediate. The 'possible' is very wide. Had the Torah permitted punishment to be carried out when the possibility is very likely – such that it is almost a necessity ... some might inflict punishment when the chances are somewhat more distant than that, and then when they are even further still, until they would punish and execute people unjustly on slight probability according to the judge's imagination. Therefore, the Almighty shut this door and commanded that no punishment be carried out except where there are witnesses who testify that the matter is established in certainty beyond any doubt, and, moreover, it cannot be explained otherwise in any manner. If we do not punish on very strong probabilities, nothing can happen other than that a sinner be freed; but if punishment be done on probability and opinion it is possible that one day we might kill an innocent man – and it is better and more desirable to free a thousand sinners, then ever to kill one innocent.[9]

It is interesting that Maimonides here emphasizes that even the testimony of witnesses can be acted upon only when the events 'cannot be explained otherwise in any manner.' This is in accordance with his view, explained and repeated elsewhere, that after all the trustworthiness of two witnesses, even when subjected to careful cross-examination, is not absolutely certain. Is there not an inherent contradiction, then, in rejecting probabilities while accepting the testimony of competent witnesses? The dilemma is solved as follows: 'For we have been *commanded* to decide the issue on the testimony of two qualified witnesses.'[10]

8 See *Mekhilta* to Exodus XXIII 7
9 Maimonides, *Sefer HaMitzvot*, Negative Commandments no.290
10 Maimonides, *Mishneh Torah: Yesodei haTorah* VIII 7: also *Sanhedrin* XXIV 1

Observe that Maimonides assumes a comparative scheme of probability in all its aspects set forth in the preceding section. Thus, let h_1 be the proposition 'The accused committed a crime,' and let the various circumstances such as those related by Rabbi Shimon ben Shetaḥ be e_1, e_2, ... Then h_1 is confirmed more strongly by the conjunction (e_1 & e_2) than by e_1 alone, and even more strongly by (e_1 & e_2 & e_3), and so on until it is 'close to certainty.' This corresponds to case (b) in Carnap's tabulation. Now Maimonides adds that, even if the probability of h_1 on the given evidence is high, that is not enough. It is also required that 'it cannot be explained otherwise in any manner.' This means that any other hypothesis h_2 is not rendered probable by any of the available evidence. In other words, he is comparing here the confirmatory effect of the evidence (e_1 & e_2 & ...) on h_1 with its effect on h_2 (case (c)) and moreover, if there exists evidence e' which is irrelevant to h_1, it must be so that h_2 is not confirmed by e' either (case (a)).

8.2 LEGAL PRESUMPTIONS AND THE PROBLEM OF INDUCTION

Among the differentia between monetary disputes and criminal proceedings, according to rabbinic law, is the fact that, in the former, evidence is admissible which renders a conclusion probable to a greater or lesser degree. Various categories of such evidences are defined in the Talmud and the rabbinic literature. These are known under the generic name *ḥazakah*, which may be translated loosely as a 'legal presumption.' In the same text where the Talmud discusses the source of the rule 'Follow the majority,' a similar question is asked with respect to *ḥazakah*. 'Whence is derived the rule which the Rabbis stated: "Place everything on its once ascertained status."'[11] This rule exemplifies the simplest kind of *ḥazakah* – the presumption, that the *status quo ante* remains as long as we have no evidence of a change. An extension of this is the presumption that a process continues in accordance with its previously established pattern. Thus, a messenger who

11 *Ḥullin* 10b

was sent to deliver a bill of divorce and left the husband old or ill, hands over the divorce to the wife, 'on the presumption that he [the husband] is alive.'[12] Another example of a process presumably persisting is 'The menses come on time.'[13]

The notion of *ḥazakah* brings into focus the problem of induction. By what right does one assume that nature displays an inertial quality maintaining the conditions of the past into the future? The problem did not escape notice. Maimonides warns not to 'categorically pronounce false any assertions whose contraries have not been demonstrated ... or that are possible, though very remotely so.'[14] Now with respect to laws of nature, he declares that their continued persistence is due to an original act of God's will which being unchangeable endows nature with uniformity.[15] 'The Divine Will ordained everything at creation so that all things will continue in accordance with their nature always, as it is said: "What has been is what will be, and what has been done is what will be done; and there is nothing new under the sun"[16] ... and of this the [sages of the Talmud] said: "The world goes its customary way." '[17] Yet, that will not suffice to vindicate the legal presumptions that the *status quo ante* has not been modified, for except in some few instances what natural laws are known to apply?

Maimonides therefore explains: 'Should we follow the possibilities, the matter would reach no end. The principle is that if

12 *Gittin* III 3
13 *Niddah* 16a
14 *Guide* I 32 (Pines, p.68)
15 *Commentary on the Mishnah: Introduction to Avot*. 8. The same idea is in *Guide* II 29 (Pines, p.346) and in *Mishneh Torah: Teshuvah* V 4. Compare J.S. Mill, *A System of Logic* Bk.III, chap.4 sec.1, p.230. See also E. Zilsel, 'The genesis of the concept of physical law,' *Philosophical Review* 51 (1942), 245–79.
The rabbinic view that physical law is ordained by God is rooted in the Bible and was apparently introduced into western thought by Philo (born *ca*. 20 B.C.E.). In his *De Opificio Mundi* (XIX) he writes about the 'operations of nature ... which are invariably carried out under ordinances and laws which God laid down in His universe as unalterable.' (*Philo*, with an English translation by F.H. Colson and G.H. Whitaker, vol.1 (London, 1929), 47).
16 Ecclesiastes 1:9
17 *Avodah Zarah* 54b

a given situation has been established we assume that it remains thus until there is a definite thing that removes that presumption, but where there is a doubt that it is possibly otherwise the presumption is not removed.'[18]

It is a very practical kind of argument Maimonides offers. Unless one accepts the principle of *ḥazakah* many legal questions will be unresolvable. Since decisions must be made, and they cannot be made on a multiplicity of contingencies, let them be based then on knowledge of the past, which is all there is to go on. However, it appears that because there seemed to be no convincing explanation, the ultimate justification was taken to be, as the Talmud expressly states,[19] a scriptural directive. This is characteristic of the rabbinic method. Scripture cannot be used to contradict reason, but where the rational lacks sanction, authority is provided by the revealed Law.

8.3 A HIERARCHY OF PRESUMPTIONS

A *status quo ḥazakah* is not the only kind of evidence of a probabilistic nature. Among other evidences are those based on statistical generalizations of different types, especially psychological insights.

Some of these other kinds of *ḥazakah* are illustrated by: 'No man sins if it is not for himself';[20] 'One does not ordinarily pay a debt before term';[21] 'No man will remain calm when his property is being robbed';[22] 'No man files suit unless he has some claim';[23] 'A man will not brazenly deny his creditor to his face.'[24]

The relative strength of 'presumptions' can be ascertained when it happens that they contradict each other. Thus the last two are opposed to each other in the following ruling where A claims

18 *Commentary on the Mishnah: Nazir* IX 2
19 *Ḥullin* 10a
20 *Bava Metzia* 5b
21 *Bava Batra* 5b
22 *Sanhedrin* 72b
23 *Shevuot* 40b
24 Ibid.

that B owes him a debt which B denies, and therefore B does not have to pay since there is no evidence.

Rabbi Naḥman said: We administer a consuetudinary oath to the defendant.
What is the reason?
[Answer] It is a presumption that no man files suit unless he has a claim.
[Objection] On the contrary, it is a presumption that a man will not brazenly deny his creditor to his face.
[Answer: This is not an instance of brazen denial] It is a delaying tactic; he thinks, '[I shall postpone the proceedings] till I get [the money] and I shall pay him.' You can see that [we admit the possibility of a delaying tactic] ... for one who denies a loan [and is proven a liar] remains acceptable as a witness, whereas one who denies that he was entrusted with an article becomes disqualified to testify.[25]

It is clear from this text that the presumption that no man files suit without cause is of wider scope than the opposing presumption. Moreover, in another passage the argument that the defendant is rationalizing his behaviour is extended even to a trustee,[26] with the result that, effectively, the former presumption is regarded as stronger.

In general, presumptions of the same type are regarded as equal and when opposed cancel each other out. However, there is a hierarchy of types. Thus Rabbi Joseph Colon (1420–1480) explains in a lengthy responsum that the first or weakest type of presumptions are those which assume the perpetuation of the *status quo ante*. They are based upon the principle traced to scripture 'that we cannot postulate new things coming into being or changes before we see them, and this we derive in *Ḥullin* [10b].' Therefore, all such presumptions are equal. However, they are essentially based on ignorance: because no current positive knowledge is accessible, nothing is admitted beyond what was known previously. Such presumptions cannot balance a

25 Ibid.
26 *Bava Metzia* 5b

ḥazakah which stems from a reason [which gives probable knowledge of the present state], such as that a man does not pay before his debt is due ... Against this it is obvious that one cannot range a *ḥazakah* of property belonging where it was, for if the former is true, the latter says nothing. If it be so that he had not paid, the debt not yet having become due, then we know that he is holding the money illegally.[27]

As between a standard majority and a *status quo* presumption, the Talmud considers the former to be the weightier[28] but certain other presumptions are taken equivalent to certainty. Such is the presumption 'that a scholar will not let go out of his hand something that is not ritually prepared.'[29] On the other hand, the tenth-century commentator Rabbi Gershom[30] emphasizes that *status quo* presumptions are based upon our knowledge of the individual and are therefore applicable even in those cases where a majority would not be, for a majority does not tell us anything about the particular individual we are interested in, whereas a presumption does. Majorities yield what might be called 'statistical probabilities.' Strictly speaking these apply to entire classes, or at least to reasonably large samples. A *status quo ḥazakah* or similar evidence gives an 'individual probability.' It follows that when a presumption is conjoined with a minority, it may increase the probability that the individual under consideration belongs to the minority. Thus the Talmudic statement: 'Join the minority with the presumption and the majority is weakened.'[31]

In general terms, in all instances where a *ḥazakah* or other evidence modifies the effect of the majority, the question at issue is that of measuring the effect of the accumulation of evidence on the prior probabilities determined before that evidence was available. This is usually handled by means of Bayes's Theorem, which was discussed in §4.5.

27 Rabbi Joseph Colon, *Responsa* no.72 (folio 29b). See also Rabbi Asher ben Yeḥiel, *Responsa*, 68:23 (p.126, col.1)
28 *Kiddushin* 80a
29 *Pesaḥim* 9a
30 Rabbi Gershom ben Judah: *Commentary to Bava Batra* 93a, s.v. סברוה
31 *Kiddushin* 80a. See the discussion in Tosafot there, s.v. סמוך. See also *Yevamot* 119b

8.4 EVIDENCE IN OTHER LEGAL SYSTEMS

The exigencies of legal procedure led also in other legal systems to a recognition of the relativity of probability with respect to the evidence, and ultimately to a gradation of probabilities. Thomas Aquinas says: 'About human acts, for which courts are set up and testimony required, there can be no demonstrative certainty ... Therefore, probable certitude suffices, which attains the truth in most cases and misses the truth in less ...'[32] Since evidence does not produce certainty, it is conceivable that different data will support contradictory hypotheses. Apparently Aquinas regarded any opposing evidences as capable of completely cancelling each other out: 'testimony ... does not have infallible certitude but is probable, and therefore anything which gives probability to the contrary renders the testimony inefficacious.'[33]

Much later, we find Leibniz, in a letter to John Bernoulli dated 6 June 1710, recalling how he was led to his ideas on probability logic from a consideration of legal probabilities. The lawyers use concepts such as 'conjecture,' 'indication,' and 'presumption,' and these correspond to different grades of probability ranging from 'not-full,' through 'half-full' to 'full.'[34] Later still, when the mathematical theory of probability had become established, it became fashionable to apply the calculus of probability to the evaluation of evidence and many different approaches were attempted in the search for a suitable quantitative concept, none of them leading to very fruitful results.[35]

32 *Summa Theologiae* II, II, 70, 2

33 Ibid., II, II, 70 3c

34 Quoted in Louis Couturat, *La Logique de Leibniz d'après des documents inédits* (Paris, 1901), p.240 n.3: 'Ego jam a puero hoc argumentum versavi, tunc imprimis cum juri darem operam, et de conjecturis, indiciis, praesumptionibus, et gradibus probationum minus plenarum, semi-plenarum, plenarum similibusque agerem.'

35 See, for example, John Tozer, 'On the measure of the force of testimony in cases of legal evidence,' *Transactions Cambridge Philosophical Society*, vol.8 (1843), 143–58

9
Induction and hypothesis

9.0 COMPETING HYPOTHESES

When specific observations or 'evidence' allow only two mutually exclusive and exhaustive hypotheses, the decision procedure is quite straightforward. As we have seen, the rabbis applied to such problems acceptance rules which determine when one of the two competing hypotheses can be effectively eliminated. The matter becomes much more complicated and difficult if there are more than two hypotheses to explain the observed evidence. We shall see that the rabbis viewed the complexities that arise as mainly of two kinds which may be summarized under the following questions.

A What kind of logically possible hypotheses need to be considered? One can dream up many explanations for given data, but is there any point in multiplying hypotheses? If not, what guideline determines when to stop manufacturing alternative hypotheses?

B Given a set of mutually exclusive hypotheses all of which seem admissible by the criterion described in A, if such exists, how do you decide between them? Clearly, if on the evidence the probability of one hypothesis h_0 is greater than 1/2, the problem may be reduced to the case of two hypotheses only, for we may lump all the hypotheses other than h_0 into one, namely 'not-h_0,' and the choice is then simply between h_0 and not-h_0. However, if no single hypothesis has probability greater than 1/2, what kind of acceptance rule will do? When the choice is among propositions none of which is much more credible than the rest, what justification can there be for selecting one over the others?

Accordingly, Maimonides concludes that in order to reject hypotheses one must 'consider how improbable they are and what is their disagreement with what exists.'[1] This is a two-pronged test involving (a) a priori plausibility and (b) explanatory power, namely the extent to which the hypothesis implies consequences that do not conform with the facts.

From his discussion it emerges that the prior probability of a proposed hypothesis may be very small, yet its explanatory power so great that it implies consequences that agree well 'with what exists,' that is, the probability of the evidence is high on the basis of the hypothesis and moreover it explains well other observations besides those that gave rise to the original investigation. Even in such a case, if there is an alternative hypothesis which can serve as well, but with a considerably higher prior probability, it should be preferred. On the other hand, one must reject a hypothesis which initially seems reasonable, but which does not entail with a sufficiently high probability the facts observed.

The examples we shall cite deal mainly with cases in which precise probability values are not assigned, but this is no hindrance since, in practice, only a comparative scheme is required for a decision.

9.1 AN ECONOMY PRINCIPLE

The Talmud proffers a maxim to limit the introduction of arbitrary hypotheses. This is stated in the course of a discussion of the incident related in the Bible about Judah and his daughter-in-law Tamar. After her successive husbands had died, Tamar connived to have a child by Judah. She disguised herself as a harlot and when Judah approached her she took as pledges his signet, cord, and

1 *Guide* II 23. Pines (p.321) translates: 'consider how great is their incongruity ...' See below p.138

staff. 'About three months later Judah was told "Tamar your daughter-in-law has played the harlot; and moreover she is with child by harlotry." '[2] When Tamar produced the pledges 'Judah acknowledged them and he said, "She is right, it is from me." ' The rabbis ask: 'How did he know [for certain]? Perhaps, just as he had come to her some other man had come to her?' ... Said Rava '... Maybe Judah had reckoned the days and months and found them to co-incide, for, what we see we may presume; but we presume not what we see not.'[3]

Here we have a modification of the principle of *ḥazakah* – in fact, the Hebrew term used here is from the same root – namely, that we do not entertain on grounds of ignorance alone alternative hypotheses, albeit they may in themselves be not improbable.

One might advance an indefinite number of competing hypotheses, all of which are capable of explaining the fact of Tamar's pregnancy: as many hypotheses as there were men in the neighbourhood. However, it is known for a fact that Judah cohabited with her at a time which, assuming that she conceived then, would explain well the present state of her pregnancy. The hypothesis that Judah is the father requires no further assumptions than that she actually conceived as a result of that union. To deny Judah's paternity would require an assumption that other men, too, consorted with Tamar, and while this may not be far-fetched in itself, it is hardly to be preferred over an explanation which is based on a known fact. Of course, if the date of the known congress were very much too late, say, to explain the present symptoms of pregnancy, one should be forced to ascribe fatherhood to someone other than Judah.

Essentially the rule 'What we see we may presume; but we presume not what we see not' is an economy principle. It precludes introducing new hypotheses, when there are old ones, established on other grounds, which are sufficient also to explain the phenomenon under investigation. It is another manifestation of the same tendency in rabbinic logic by virtue of which it is postulated that

2 Genesis 38:24
3 *Makkot* 23b

there are no redundancies in the revealed text of the law.[4]

It is instructive to compare another instance of rabbinic reasoning in which the same principle is applied differently. There was a controversy about the pace of development of the human embryo. Rabbi Ishmael argued against most of the sages that, while a male foetus is already recognizable as such at 41 days after conception, a female foetus does not reach that stage until 81 days. The other rabbis held that there is no difference in this respect between male and female.

> Rabbi Ishmael said to them: 'It happened to Cleopatra, a Greek queen, that her maidservants were condemned to death. [and so the authorities experimented with them by having them impregnated and then isolated in prison]. They examined them [post-mortem] and found a male foetus [formed] at forty-one days and a female at eighty-one [at a similar stage of development].'
>
> They [the sages] said to him [Rabbi Ishmael]: 'One does not bring a proof from fools!' What is the reason [of their objection]? The one bearing the female delayed and then conceived later ... there is no guardian for lust and I may suppose that the [prison] guard consorted with her.[5]

In this case, as in the one involving Judah and Tamar, there was a known instance of coition. If that was the cause of the pregnancy, the observed results would indeed suggest that the sexual difference might be the reason for the disparity in development. In the case of Tamar, the hypothesis that other men consorted with her besides the known one was dismissed summarily. The very same hypothesis is invoked in the case of Cleopatra's maids, despite the fact that in prison not many men were available. Rather than accept what was to them an implausible hypothesis that there is a great difference in the rate of maturation between male and female foetuses, the rabbis were willing to accept even 'what we see not.'

4 See above p.12. This principle is similar to the so-called *Law of Parsimony* of the Latin scholastics or, as it is sometimes known, *Ockham's razor*, that multiplicity ought not to be posited without necessity.

5 *Niddah* 30a,b

The distinction between the two cases is not difficult to find. In the one case, to deny Judah's paternity is less credible than to assert it. In the other case, to posit that the jailer consorted with the maids is far more plausible than to suppose that there is a radical difference between male and female foetuses.[6] To establish such a general law, much stronger and repeated proofs would be required. The choice is between (a) a restriction to the known occurrence of intercourse (*I*) which necessitates admitting an unknown law of physical development (*D*), and (b) admission of the possibility of later intercourse (*L*) which allows retention of the assumed similarity of male and female foetal growth (*G*). Thus the principle that 'what we see we may presume; but we presume not what we see not' is selectively applied. In fact, we have here a kind of significance test whereby the relative probabilities enable one to determine which of the conjunctions (*I* & *D*), (*L* & *G*) is to be rejected.

9.2 ARE THE STARS RANDOMLY DISTRIBUTED?

An example from a much later period will illustrate the further development and application of the same procedure, whereby hypotheses are evaluated in terms of their prior probability as well as by their explanatory power.

Maimonides[7] argues that the universe is the handiwork of an intelligent purposive creator. He claims that his proofs, while only

6 Apparently the view that female foetuses develop more slowly was widely held. Aristotle says: 'While within the mother the female takes longer in developing' (*De Generatione Animalium* IV 6, 775a12). Presumably under Aristotle's influence the same doctrine was taught in the Middle Ages. Thus Rabbi Menaḥem Meiri reports that 'most natural scientists attest to Rabbi Ishmael's opinion that the form of the female is not complete till eighty days' (Commentary to *Niddah* 30a). Rabbi Gershon ben Shlomo (thirteenth century) in his Encyclopaedia entitled *Shaar haShamayim* cites a compromise view that equalizes the stage reached by males at forty days to that attained by females at sixty days (chap.8, p.49).

7 *Guide of the Perplexed* II 19 (Pines, p.309–10). I have paraphrased much of the discussion, quoting only the most salient passages. The principle of 'particularization' is stated in *Guide* I 74 (Pines, p.220). See above § 5.5

inductive, when taken together show that this conclusion has such high probability as to make it 'near demonstrative certainty.' Only one of the arguments will be cited here. It is entirely based on the principle of 'particularization' which Maimonides proposed in order to discriminate between phenomena due to chance and those due to design. 'There can be giving of preponderance and particularization only with respect to a particular existent that is equally receptive of two contraries or of two different things. Accordingly it can be said of that, inasmuch as we have found it in a certain state and not in another, there is proof of the existence of an artificer possessing purpose.' If two possibilities are equally likely, and one is observed to predominate, the phenomenon must be due to a purposeful maker. He reasons from the observed distribution of the stars in the heavens. The possible hypotheses are three.

1 The distribution is due to chance.

2 The Aristotelian view that this 'proceeded obligatorily and of necessity from the Deity' in some unknown manner implies the existence of unsuspected natural laws, which account for the observed irregular and asymmetric distribution.[8]

3. It is due to design.

As for the first alternative, Maimonides says:

> It is even stranger that there should exist the numerous stars that are in the eighth sphere, all of which are globes, some of them small and some big, one star being here and another at a cubits' distance according to what seems to the eye, or ten stars being crowded and assembled together while there may be a very great stretch in which nothing is to be found. What is the cause that has particularized one stretch in such a way that ten stars should be found in it and has particularized another stretch in such a way that no star should be found in it? ... the whole sphere is one simple body in which there are no differences ... How then can one who uses his intellect imagine that the positions, measures and numbers of the stars ... are fortuitous?

8 Maimonides remarks that Aristotle does not explicitly state this, but it is a consequence of his system as expounded by the followers of his school. *Guide of the Perplexed* II 13 (Pines, p.284)

Aristotle[9] had argued that regularity and symmetry preclude chance and indicate natural law, but here we have neither regularity nor yet random distribution.

If chance is rejected, there remain either unknown necessity or design for some unknown purpose. Now purposive design is known as an efficacious cause for irregularities.

In fact you know that the veins and nerves of any individual dog or ass have not happened fortuitously, nor are their measures fortuitous. Neither is it by chance that one vein is thick and another thin, that one nerve has many ramifications and another is not thus ramified, that one descends straight down and another is bent. All this is as it is with a view to useful effects whose necessity is known.

Is it reasonable then to postulate unknown natural necessity to explain the strange distribution of the stars, which is clearly neither random nor regular, when there is a familiar principle at work in the world which can explain the irregularity? That would be 'very remote indeed from being conceivable.' On the other hand though, 'If it is believed that all this came about in virtue of the purpose of one who purposed who made it thus, that opinion would not be accompanied by a feeling of astonishment and would not be at all unlikely. And there would remain no other point to be investigated except what is the cause for this having been purposed?'

The subsequent history of this Maimonidean argument is interesting indeed. In 1767, John Michell[10] examined the hypothesis that the stars are distributed at random over the celestial sphere, and he tried to calculate the probability on this hypothesis of getting a constellation like the Pleiades where six stars cluster together. Sir Ronald Fisher has reworked the computation and finds that the probability 'is amply low enough to exclude at a high level of significance any theory involving a random distribution.'[11]

9 *Physica* II 4, 196b10

10 'An inquiry into the probable parallax and magnitude of the fixed stars etc. ... ' *Transactions Royal Society of London* (1767), vol.57, XXVII (abridged ed.v.12 [1809], 430 ff.). See also J.M. Keynes, *A Treatise on Probability*, pp.294 ff.

11 R.S. Fisher, *Statistical Methods and Scientific Inference* (London, 1956) p.39. Probabilistic proofs for design in creation were given by Newton and others. See O.B. Sheynin, 'Newton & the Classical Theory of Probability,' *Archive for*

9.3 FALSIFYING HYPOTHESES

As we have seen in our second Talmudic example, the scope of the principle 'what we see we may presume: but we presume not what we see not' is limited by the important requirement that the known hypothesis is to be rejected if it leads to results which appear strange and improbable. In that case, even an unknown hypothesis may be more acceptable. One final example from the Middle Ages will show how this consideration worked out in practice. Before citing this illustration, we shall digress for a moment to consider some relevant modern ideas.

Recent writers on the philosophy of science assign a central role to the *hypothetico-deductive method*. In this view 'the work of the scientist consists in putting forward and testing theories.'[12] As to the origination of new theories, this is regarded as more or less beyond the province of strict logical analysis, since in the formulation of any significant hypothesis there is usually a factor of inspiration or intuition at work. It is in the testing of theories that the rules of scientific method find application. Implications that can be derived by deduction from the proposed theory are put to experimental test. The scientist proceeds by trying to falsify the theory, that is to say, he attempts to draw deductive conclusions from the theory in conjunction with the rest of the body of accepted knowledge – conclusions that are predictions of an empirical nature. As such they can be tested by setting up suitable experiments. If the predictions of the theory fail to be corroborated by repeated experiments, the falsification of these predictive implications of the theory is considered to falsify the theory itself.

We shall see that a scientist in the Middle Ages worked in much the same way. The matter under investigation was the association of heat with sunlight. Is heat a property of light or is the heat generated in some other way?

On the question of how heating is effected, Aristotle had already noted that friction causes heat.

History of Exact Sciences 7 (1971), 217–43. For evidence that Newton was familiar with and was influenced by Jewish teachings, see Max Jammer, *Concepts of Space* (Harvard University Press, 1954), pp.108 ff.

12 K.R. Popper, *The Logic of Scientific Discovery*, p.31

The warmth and light which proceed from them [the sun, moon and stars] are caused by the friction set up in the air by their motion. Movement tends to create fire in wood, stone, and iron, and with even more reason should it have that effect on air, a substance which is closer to fire than these. An example is that of missiles, which as they move are themselves fired so strongly that leaden balls are melted; and if they are fired the surrounding air must be similarly affected.[13]

Thus not only is locomotion a prerequisite for heating to take place: it actually generates heat itself.

Although he agreed that friction generates heat, Rabbi Levi ben Gershon (1288–1344) argued against Aristotle's explanation of the sun's heat as due in large part to its motion and the resultant friction in the air. Among other objections, he points out that 'the motion of the sun ... cannot be transmitted to the elements, but [its effect] ends and ceases at the matter which is between the spheres of the sun and the spheres of the next planet below it.'[14]

Therefore, Rabbi Levi ben Gershon picked up another of Aristotle's suggestions, namely, that light has a specific power to cause heating. In examining this hypothesis, he deduces some of its implications, and puts them to experimental test. On this basis he finds that some kinds of light, for example moonlight, do not have heating power, but sunlight does. Moreover, he finds that sunlight acts in the same way as the light from a candle so that there must be something more to the rays of sunlight or candle-light than just light. What this power of sunlight is, he cannot say except that it increases with the concentration of the light. He is driven therefore to posit an unknown 'divine power,' which like so many other works of the Creator 'is not completely comprehended by any intellect besides Him.' The argument is summed up as follows.

13 *De Caelo* II 7, 289a20

14 Rabbi Levi ben Gershon, *Milḥamot Hashem* V, part 2, chap.6. Levi made other more sophisticated observations and experiments in order to falsify astronomical hypotheses and in this manner he rejected epicycles in favour of eccentrics. See Bernard R. Goldstein, 'Theory and Observation in Medieval Astronomy,' *Isis* 63 (1972), 39–47

A ray of sunlight, by virtue of a relationship between it and fire, with a divine power that it possesses moves fire and heats the air in this way due to the admixture of fire in it [the air]. Therefore, since on being reflected the ray is multiplied on itself to a remarkable degree, so too the heat is multiplied when the reflections of the ray are greater ...

Proof for this can be seen in 'burning mirrors' – many rays are reflected and co-incide at one point and that is the place where burning occurs. Since this is so, this is a property of light qua sunlight but not insofar as it is absolute light. Because this is found in sunlight by virtue of the nature of the element fire which the sun's ray moves, this heating property ought to be found also in light from fire when it is reflected. Now that is something we have established from sense-experience. We placed a 'burning mirror' opposite the light of a candle and at the point where the reflected rays coincide we found that it burns. But for moonlight you will not find burning created by a 'burning mirror.'

9.4 A RABBINIC PRINCIPLE OF MAXIMUM LIKELIHOOD

When the probability of none of the competing hypotheses is greater than 1/2, must they all be rejected? If that be so, no decision could be made and judgment would be suspended. If the competing hypotheses are all equally probable, we have already seen that the Talmud regards them all as inconclusive and Maimonides explains that even if one of equipossible alternatives turns out to be true, no conclusions can be drawn from this fact.[15] However, if the probabilities are not equal, the less likely hypotheses may be rejected, leaving the one of maximum likelihood as the only and therefore acceptable conclusion. This far-reaching guideline is set forth by Rabbi Asher ben Yeḥiel in his analysis of a Talmudic precedent.

The Talmud tells of a man who learned that of the ten sons his wife bore, he was the father of only one. Before his death, he bequeathed his entire estate to his only son, without specifying

15 See above p.103

who he was. The case came before Rabbi Bana'ah who instructed the ten claimants, 'go beat on your father's grave until he will arise and reveal to you to whom he left his [property].'[16] Only one of the sons refrained from such disrespectful behaviour, and so Rabbi Bana'ah awarded the bequest to him. Rabbi Asher in the course of a responsum dealing with verdicts based on probabilities comments:

> On the strength of this [case of Rabbi Bana'ah], in an issue which can be determined, we do not say 'Let it stay till Elijah comes' [that is – let judgment be suspended and the property be held in escrow by the court]: rather let him judge in accordance with what his eyes behold, a probable opinion. On a small probability – that it appeared to him that the true son had respect for his father firmly implanted in his heart and paid him honour – he gave him all the property.[17]

Here we have a case where apparently all the claims are initially of equal probability. The test devised to ascertain who is the true son is certainly not a very decisive one: Rabbi Asher emphasizes that it is only a 'small probability.' Yet because one claim has a higher probability than any of the others, we are justified in reaching a decision.

It ought to be pointed out that this case is quite unlike that discussed above in § 5.6[18] where the ruling is that 'property that is in doubt is divided equally.' The Talmud distinguishes between a dispute where all the conflicting claims have equal probability but only one has true validity, and a case in which all claims may in fact be equally valid, so that an equal division need not necessarily be in error. In the latter case, equal division is indicated, but in the former, since the entire property really belongs to one litigant only, if the evidence is equally balanced, no verdict can be rendered and the property must be held in escrow 'till Elijah comes.'[19]

It is noteworthy that Keynes cites a proposal by Leibniz that

16 *Bava Batra* 58a
17 Rabbi Asher ben Yeḥiel, *Responsa* 107:6 (p.196, col.2)
18 See p.76
19 *Bava Metzia* 3a

if two people put forward claims to a sum of money which can belong to one only, one claim being twice as probable as the other, the sum should be divided in that proportion between them. Keynes himself comments: 'The doctrine seems sensible.'[20] According to the Talmudic concept of equity, it is neither sensible nor just. Since the entire sum belongs to only one, the rabbis would assign the complete amount to him whose claim seems best founded.[21]

9.5 THE MODERN PRINCIPLE OF MAXIMUM LIKELIHOOD

In modern statistical theory, an important principle of inference is the 'maximum likelihood principle.'[22] A particular state of nature θ_i is thought of as an explanation of the observed data, say the value of a variable X. The probability of obtaining the observed value depends upon the state of nature which is assumed to govern. Thus, if one gets a reading for the variable $X = k$, say, one computes the probability of getting this reading on the assumption of θ_1, θ_2, etc. The probability depends upon the value k as well as on which state of nature is assumed. This defines the likelihood function $L(\theta)$.

$$L(\theta) = p(X = k|\theta).$$

$L(\theta)$ is a function of two variables: k, which is the observed datum, and θ, which describes the state of nature.

Having observed a particular value $X = k$, allow θ to wander over all possible states of nature and select that value of θ, say $\bar{\theta}$, which maximizes the probability $L(\theta)$ of obtaining the result actually observed. $\bar{\theta}$ gives the best explanation of the experimental datum. The

20 *A Treatise in Probability*, p.311, n.1

21 See Chaim Perelman and L. Olbrechts-Tyteca, *The New Rhetoric* (translated by J. Wilkinson and P. Weaver, University of Notre Dame Press, 1969), p.258, who take exception to Keynes's position on similar grounds

22 See B.W. Lindgren, *Statistical Theory* (Macmillan, 1962), pp.188 ff. Also B.W. Lindgren and G.W. McElrath, *Introduction to Probability and Statistics* 2nd ed. (Macmillan, 1966), pp.84 ff. The principle in a form suited to modern statistical theory was advanced by Ronald A. Fisher. See his *Contributions to Mathematical Statistics* (New York, 1950), especially 10.310 and 10.323

Principle of Maximum Likelihood calls for acting on the assumption that the actual state of nature is the one that gives the best explanation of the facts.

In many cases the 'best' explanation may not be good enough. In other words, the state of nature $\bar{\theta}$ for which $L(\bar{\theta})$ is a maximum may not be much more likely than the other possible states, and acting as if $\bar{\theta}$ is the true state may involve the possibility of costly error. One defines, therefore, for any two alternative states of nature θ_1 and θ_2 the *likelihood ratio*,

$$\lambda = L(\theta_1)/L(\theta_2).$$

Now for $\bar{\theta}$ and any other θ_i, $\lambda = L(\bar{\theta})/L(\theta_i) > 1$, since $\bar{\theta}$ is the most likely state of nature. In certain types of decision problems, one might require that to justify acting as if $\bar{\theta}$ is the true state, the likelihood ratio for $\bar{\theta}$ and any other θ_i must exceed one by at least some fixed number. A decision rule can be specified by setting a critical value of λ such that if λ is greater than this critical value, act as if $\bar{\theta}$ is true; otherwise do not.

A majority acceptance rule is, of course, an application of the Principle of Maximum Likelihood, and the concept of significance level can be understood in terms of the likelihood ratio. The real importance of the maximum likelihood principle is that it can be used to draw reasonable inferences even when the maximum likelihood is less than 1/2. As shown above, the rabbis had an elementary maximum likelihood principle of their own which is based on this idea.

9.6 A LIKELIHOOD RATIO

A more precise application of the Principle of Maximum Likelihood in the Talmud is made in a ruling dealing with the admixture of objects used in idolatrous worship. So grave is the prohibition of idolatry that 'even one in ten thousand' disqualifies the entire mixture. The Talmud[23] quotes a statement of Rav: 'If an idol's ring was mixed up with [others making a total of] one hundred rings and

23 *Zevaḥim* 74a

forty of them were separated to one place and sixty to another ... the forty detached to one place do not render others forbidden: the sixty in one place do render others forbidden.'

Although the odds are 3:2 that the idol's ring is in the sixty, this is not sufficient reason to permit drawing from the forty, since on an individual draw from either group the probability is the same (1/100) that the forbidden ring is chosen. Thus the remark: 'Why is one from the forty different? [Presumably] because we say, "The forbidden one is among the majority" [namely, the sixty]. Then in the case of one from sixty too we must say, "The forbidden one is in the majority" [i.e., the remaining fifty-nine]!' Both possible arguments based on 'Follow the Majority' are turned down because the likelihood is the same for any single ring in either group that it is the idol's ring. Therefore, as long as only those from the original hundred are present, none of the rings can be permitted. However, if both the group of forty and that of sixty are each mixed with other rings, the probability that a single ring in either mixture is the forbidden one is no longer the same for both groups. In such a case, the ruling is that the new mixture derived from the group of forty is permitted while that containing the group of sixty is prohibited. For example, suppose that one additional ring is added to the forty and one to the sixty, and so one obtains two sets of rings containing 41 and 61 rings respectively. The probability that a single ring of the 41 is the idol's ring is given by $40/100 \times 1/41$, while the probability of drawing the forbidden ring from the 61 is $60/100 \times 1/61$, and the latter number is greater than the former.

While no computations are given in the text, it is remarkable that the division of 40:60 is chosen rather than 49:51 which would make a stronger point.[24] However, a calculation of the different probabilities is revealing, and we can attempt to reconstruct what might be a plausible line of Talmudic reasoning, although it is mere speculation whether the analysis was carried so far.

Suppose to the set of x rings ($x < 50$), separated from the original mixture, is added at least one more ring (for the worst case)

24 See Rabbi Shlomo ben Adret, *Torat HaBayit HaArokh* IV 2 (74a)

to make the set A. Similarly, to those remaining from the original mixture, add at least one to form the set B.

The probability p_A of drawing the idol's ring from the set A is given by

$$p_A = \frac{x}{100} \cdot \frac{1}{x+1}, \quad \text{while} \quad p_B = \frac{100-x}{100} \cdot \frac{1}{101-x}.$$

Clearly, $p_B/p_A > 1$.

In other words, the *likelihood* that an arbitrary ring drawn from A (or B) is the idol's ring is given by

$$L(A) = p_A = \frac{x}{100} \cdot \frac{1}{x+1} \text{ and } L(B) = p_B = \frac{100-x}{100} \cdot \frac{1}{101-x}.$$

The *likelihood ratio* $\lambda(x)$ is the ratio $L(B)/L(A) = p_B/p_A$; it follows that $\lambda(x) > 1$.

It is necessary then to ascertain the *critical value* of λ. Thus, $\lambda - 1 = (p_B - p_A)/p_A$ can be considered as an indicator of the greater likelihood that a single draw from B, rather than from A, gives the forbidden ring, and this must be of significant magnitude. We seek a lower bound for this quantity.

Now for large x, both p_A and p_B are close to $1/100$; and since one in ten thousand renders a mixture unfit, a probability of one in ten thousand relative to one in a hundred is still decisive, i.e., the set A can be permitted if $(\lambda - 1)$ is near

$$\frac{1}{10{,}000} \Big/ \frac{1}{100} = \frac{1}{100}.$$

In fact, the minimum significant probability must be somewhat less than one in ten thousand. Therefore, the upper bound for x should not be the least value such that $\lambda(x) - 1 < 1/100$. It suffices to choose the next larger x.

We get the following table:

x	38	39	40
$(\lambda(x) - 1)$	1/99.75	11/1209 = 1/109.91	1/122

Therefore $x = 40$ is chosen as the maximum admissible value for which the set A can be permitted.

10
Subjective probabilities

10.0 SUBJECTIVISTIC PROBABILITY

Perhaps the most recent interpretation of probability is the subjectivist. On this view, probability refers to the individual's belief, with the important proviso that the individual concerned be a rational person. F.P. Ramsay explains that 'the degree of a belief is a causal property of it, which we can express vaguely as the extent to which we are prepared to act on it. This is a generalization of the well-known view, that the differentia of belief lies in its causal efficacy.'[1]

The qualification imposed by the requirement of rationality means simply this. There are people who believe and act upon propositions that are mutually contradictory or demonstrably false. Obviously, their set of beliefs as a whole cannot be taken as rational. Holding a belief in one particular proposition immediately imposes 'rational' restrictions upon which other propositions may be believed by the same person. For example, the contrary proposition is excluded.

This extends also to degrees of belief. Thus, suppose that a man has a strong suspicion that a given die is biased. In fact, he is willing

1 Frank P. Ramsay 'Truth and probability' reprinted in *Studies in Subjective Probability*, ed. H.E. Kyburg and H.E. Smokler (New York: Wiley, 1964), p.71. This essay first appeared in 1931.

to bet 3:1 odds that it will turn up a six. Holding this belief of degree 3/4, he is restricted from taking on odds of 1:1, say, that the same die will turn up any face other than a six; to be rational, the only bet he could make would be to stake one against an opponent's three, indicating a belief of degree 1/4. In this way, if he made both bets, one on placing three against one that a six will be cast, and the second staking only one against three that one of the other faces will turn up, he would neither lose nor gain regardless of the outcome. For, if a six turned up, he would collect one and pay out one, and in the alternative case he would collect three and pay out three. In other words, a necessary and sufficient condition to prevent finding himself in a position where a book can be made against him on a single event E, is that his degrees of belief in E and in the negation of E add up to one. In general it can be shown that if his degrees of belief satisfy the axioms of the probability calculus, no book can be made against him. This is also a necessary condition. Any set of degrees of beliefs which satisfies the probability axioms is called *coherent*.

It appears then that just as the requirement for consistency is a logical limitation on the colligation of propositions without casting light on the objective truth of either of them, so too coherence is a criterion which determines when a set of beliefs is admissible, without offering any guidance as to whether any of them correspond in any sense to reality. The ultimate choice of what degree of confidence is warranted in this or that proposition is purely subjective as long as it is understood that empirical data are always included in the set of one's beliefs, and they must therefore be taken into account in determining whether or not one's beliefs are coherent.[2]

I have not found a rabbinic writer who would go quite so far in giving probability a subjective interpretation. In all the sources known to me, it is assumed, at least implicitly, that whenever a probability measure or even just an ordering can be assigned, this is meant to describe in a unique way an objective reality. However, the actual process of assigning probabilities to events, say, cannot

2 See Henry E. Kyburg Jr., *Probability and Inductive Logic*, p.70

always be carried out in a manner that is clearly interpersonal and independent of any particular observer.

Thus far, we have considered probability assignments derived from types of evidence that can be objectively defined. To use Kneale's[3] term, we might say, that in the rabbinic view, assorted evidences 'probabilify' legal conclusions in accordance with a specified scale which is presumably established on rational criteria that can be accepted by all. However, there is a large class of evidence which can properly be conceived of as 'probabilifying' or lending credibility, but which, the rabbis felt, does not permit classification on an objective basis. Yet to exclude this from consideration would lead in many instances to manifest injustice. The making of an estimate of probability may in some instances be entirely subjective, being based upon indications so obscure that they defy description and analysis and are properly called hunches.

10.1 HUNCHES AND INTUITION

The matter is best summarized in the words of Maimonides in his Code:[4]

> The judge may decide in monetary cases on the basis of things which in his opinion are true and he is strongly convinced that it is so, even though there is no clear proof. It goes without saying that, if he knows for certain that the matter is so, that he can judge in accordance with what he knows. How is this? Suppose a man [defendant] was liable to an oath administered by the court, and the judge is informed by one whom he considers trustworthy and on whose words he relies, that that man is suspect for perjury – the judge may reverse the oath administering it to the claimant who can then collect on the strength of his oath. That is so because in the judge's opinion the words of this one [informant] are reliable ... Since he finds himself firmly convinced – he can rely on it and judge accordingly. It goes without saying that this is so if he [the judge] himself knows the defendant to be suspect. Similarly, if a note of obligation appears before him and he

3 William Kneale, *Probability and Induction*, p.11
4 *Mishneh Torah: Sanhedrin* XXIV:1, 3

is told by someone whom he relies upon, even ... a relative [the testimony of relatives is not admissible in court]: 'That is paid' – if he trusts his words – he may say to this [claimant]: 'You cannot collect without an oath' ... Similarly, if the claimant argues that he entrusted something to one who has since died without instructing his heirs, and he [the claimant] gives unmistakable identification marks [for the claimed object], and the claimant was not wont to frequent the dead man's house [where he might have observed his possessions] – if the judge knows that owning such an object was incommensurate with the dead man's status, and he is convinced that the object is not the dead man's – he may remove it from the custody of the heirs and give it to this one who is apt to own such and gave identification. So too, in any analogous situation – for the matter is exclusively given over to the heart of the judge in accordance with what appears to him to be the truth. If so, why did the Torah require two witnesses? That if two witnesses appear before the judge – he shall decide in accordance with their testimony even if he does not know whether they testified truly or falsely ...

Whence do we derive that if a judge knows that a case is fraudulent he may not say: 'I shall decide it and let the responsibility rest with the witnesses?' It is written: 'Keep far from falsehood.'[5] What shall he do then? He must investigate and cross-examine as one does in capital cases: if it then appears, in his opinion, that there is no fraud – he decides the case in accordance with the testimony. However, if his heart is troubled that there is fraud; or he does not trust the witnesses even though he is unable to disqualify them; or if his mind is inclined that this litigant is a fraud and a cheat who influenced the witnesses, though they be honest and testified innocently, but he misled them; or if it seems to him [the judge] from the general scheme of things that there are other hidden matters which they do not wish to reveal – in all such cases it is forbidden to issue a verdict ... Now these matters are given to the heart [and God alone knows whether the judge is ignoring his intuitions] and scripture says: 'For judgment is the Lord's.'[6]

Although it is clearly acknowledged that subjective feelings

5 Exodus 23:7
6 Deuteronomy 1:17

have an important place in deriving probability judgments, so much so that even the validity of competent witnesses can be placed in question on purely subjective or intuitive grounds, I have not found any attempt by the rabbis to formulate scales or standards of measurement. In fact, Maimonides' reference to the verse 'For judgment is the Lord's' seems to indicate that there can be no interpersonal criteria for specifying what warrants intensity of conviction and its concomitant subjective probability. Subjective probability can result, as we have seen, from very vague and inchoate stimuli which can hardly be described as evidence, or it may be generated by more rational considerations. Since the only measure of the relative validity of these indicators is subjective feeling, Maimonides stresses that it is supremely important to approach each question with an open mind unbiased by prejudice and undisciplined habits of thought. One must be constantly conscious of the awesome responsibility incumbent upon the sincere seeker of truth, who is answerable to the Almighty Himself.

10.2 SUBJECTIVE OBJECTIVITY

In another context, Maimonides develops the same thought, intimating that as we range from the least substantial shreds or suggestions of evidence to those whose probabilifying value can be more nearly objectively ascertained, the reliability of our feelings as arbiters of probability can be tested. The matter is discussed in connection with a philosophical problem in the *Guide of the Perplexed.*

One of the most vexing difficulties which confronted all those attempting to synthesize Aristotelian philosophy with Biblical faith was the question of creation. According to Aristotle the world is eternal and uncreated. The Bible, however, teaches that God created all that is. Here was an open and unbridgeable contradiction that could only be resolved by rejecting one or the other alternative. Yet, if Aristotle's view were scientifically proven, its denial would amount to a rejection of reason. Maimonides argues that the eternity of the world has not been demonstrated. He goes much further and declares that it is, in principle, undemonstrable. And therefore, of course, so is the alternative.[7] Consequently, he attempts to clarify the role of evidence in the absence of demonstration. He makes it clear that evidence is weighted, sometimes subjectively, and that the resulting probability of propositions based on this evidence often cannot be measured in numerical terms. Yet, a comparative scheme can sometimes be determined, and when personal preferences or inclinations can be overcome, a near objective evaluation of the probability of the conclusions can be reached. This requires that all the arguments, pro and con, inconclusive though they may be, be scrutinized and weighed. While demonstrative certainty cannot be attained, a greater or lesser degree of confidence in the conclusion can be justified. Evidence is considered to be any consideration that appears capable of influencing opinion.

Know that when one compares the doubts attaching to a certain opinion with those attaching to the contrary opinion and has to decide which of them arouses fewer doubts, one should not take into account the number of the doubts but rather consider how great is their incongruity[8] and what is their disagreement with what exists. Sometimes a single doubt is more powerful than a thousand other doubts. Furthermore this comparison can

7 *Guide* II 23. In Pines's English translation: '... a demonstration [that the world is eternal] does not exist in nature' (p.322). Compare the Talmudic logical concept of *teku*, mentioned above p.20. Maimonides' position on this question deeply influenced the Christian scholastics. See Etienne Gilson, *Le Thomisme* (Paris, 1923), p.111

8 The Arabic word is שנאעתהא. The Hebrew has here the usual medieval word for 'improbability': הרחקתם, literally remoteness (see below p. 157). Friedlander translates the phrase 'the degree of improbability' (M. Friedlander, *The Guide for the Perplexed by Moses Maimonides*, 2nd ed., London, 1936, p.135). So also Munk, translating into French renders the word 'l'invraisemblance' (S. Munk, *Le Guide des Égarés – Maimonide*, vol.2, Paris, 1861, 181). The Latin version made in the early thirteenth century reads: 'Sed in magnitudine convenientis quod sequit: vel inconvenientis quod sequit,' which seems to be an interpretation in which the following phrase 'and what is their disagreement with what exists' is telescoped together with the first condition, 'how great is their incongruity' (*Moses Maimonides ed. Augustinus Justinianus Parisiis 1520*, Minerva G.M.B.H. [photo-offset], Frankfurt a.M., 1964).

Pines's word 'incongruity' corresponds best to the Latin. In any case the sense is clearly one of lack of plausibility or credibility

be correctly made only by someone for whom the two contraries are equal. But whoever prefers one of the two opinions because of his upbringing or for some advantage, is blind to the truth. While one who entertains an unfounded predilection cannot make himself oppose a matter susceptible of demonstration, in matters like those under discussion [namely, probabilities], such an opposition is often possible. Sometimes, if you wish it, you can rid yourself of an unfounded predilection, free yourself of what is habitual, rely solely on speculation, and prefer the opinion that you ought to prefer. However, to do this you must fulfil several conditions. The first of them is that you should know how good your mind is and that your inborn disposition is sound. This becomes clear to you through training in all the mathematical sciences and through grasp of the rules of logic ... [9]

10.3 DEGREES OF BELIEF

Is there an independent measure of conviction? In other words, can a person's degree of belief be determined by criteria outside of his own feelings? The rabbis regarded probability assignments, even when made on the basis of highly subjective intuition, as attempts to approximate an objective probability value – as Maimonides puts it: there is an opinion which 'you ought to prefer' and that 'ought' reflects objective probability. It is, therefore, necessary to distinguish between belief and probability, for the latter is a datum of knowledge while the former represents an emotional involvement and commitment to the thing known. Of course, Maimonides maintains that the *subjective* degree of belief ought to depend upon the *objective* degree of probability. Yet the relation need not be one of strict proportionality, for that is not the only factor to be considered. The decision to believe, no less than any other decision, will be affected by personal inclinations and values. What are the gains to be expected as well as the losses to be guarded against, and how significant are they for the particular individual?

Rabbi Levi ben Gershon emphasizes that even for empirical sense-data confidence is warranted only after repeated observa-

9 *Guide* II 23; Pines, p.321

tions although the content of the information obtained each time is the same. Granting the claims of astrology that the stars are determining factors in our lives, a true understanding of the effects of the celestial bodies is beyond us, Rabbi Levi claims, because, among other reasons, 'it cannot be that our observations will be repeated as many times as is necessary for experimental conclusions of this kind. For the positions of the stars in ascendancy cannot recur before many thousands of years multiplied exceedingly many times.'[10] Thus belief grows with experience, but that growth represents more than just an increase in the probability of the data. What other long-standing beliefs are threatened by admitting a new one? What reshuffling of views and even basic outlook is imposed by accepting a given conclusion? To have observed a particular fact several times may well be enough to convince even the most sceptical that it is so, but to believe that the predictions of astrology are valid and represent universal laws is a matter of such moment that only very many repeated and varied observations could provide an adequate foundation for such a belief.

The rabbinic outlook on probability is, as we have seen, entirely decision oriented. It is therefore to be expected that, if beliefs or convictions are to be measured, it must be in terms of readiness to act on them. Thus it is not surprising that already in the Bible absolute belief in God is equated with readiness to sacrifice even life itself to defend this belief. Thus the command: 'Hear O Israel, the Lord is our God, the Lord is one: You shall love the Lord your God, with all your heart and with all your soul and with all your might.'[11] The Talmud comments: '*With all your soul* – even if he takes your soul.'[12]

In the middle ages, Jewish thinkers distinguished between belief

10 *Milḥamot ha Shem* II 2. See Edward Grant 'Nicole Oresme and the commensurability or incommensurability of the celestial motions,' *Archive for History of Exact Sciences*, 1 (1961), 440 for a reference to Oresme's earlier view that 'the period will be much greater than 36,000 years.' This may be another indication of a possible dependence upon Rabbi Levi which is discussed below, p.150

11 Deuteronomy 6:14

12 *Berakhot* 61b

on the strength of tradition and belief that is generated by knowledge.[13] The former involves committing oneself to the truth of certain propositions and acting on them, even though they are not directly intuited and one knows neither empirical nor rational evidence entailing these truths, other than the testimony of respected authority. On the other hand, belief which results from knowledge can come about only as a result of direct experience or rational evaluation of evidence. However, in practice, the intensity of both kinds of belief is measured in the same way – by the behaviour which they inspire. In a given situation, the risks one is prepared to take in acting upon the truth of a given proposition depend upon the gravity of the consequences upon failure to act and the degree of one's belief in the truth of the said proposition.

Kneale[14] suggests the use of three separate terms to distinguish between these three kinds of belief. 'Taking for granted' describes that state of mind that accepts a conclusion without thinking it through independently. '*Taking for granted* is a kind of behaviour rather than a kind of thought ... it consists in behaving as though one knew something which one does not in fact know.' Belief based on rational considerations without certain proof is called 'opinion,' while that which is demonstrated is 'knowledge.'

13 Maimonides, *Guide* III 54 (Pines, p.633). See also the commentary of Rabbi Levi ben Gershon to Deuteronomy 6:4

14 William Kneale, *Probability and Induction*, pp.5–6

11
Combinations and permutations

11.0 THE PURE MATHEMATICS OF PROBABILITY

The subject of combinations and permutations is very intimately connected with probability. Without their help the enumeration of equally possible cases cannot be accomplished, other than in the very simplest kinds of problems.

Today, one conceives of the mathematical theory of probability in purely formal axiomatic terms as a branch of measure theory, and the interpretation which connects the mathematical theory to reality is entirely extra-mathematical. The assignment of probabilities to events is solely a philosophical undertaking. Strictly speaking, for the mathematician it is irrelevant whether there is any empirical or other correlate to his probability functions. In much the same way John Venn thought of the elementary theory of his day when he wrote: 'When Probability is ... divorced from direct reference to objects, as it substantially is by not being founded on experience, it simply resolves itself into the common algebraical or arithmetical doctrine of Permutations and Combinations.'[1]

The problems which are dealt with in the rabbinic literature are all, as we have seen, quite easily handled even without formal methods, for the possibilities can be quite readily enumerated.

1 John Venn, *The Logic of Chance*, 3rd ed. (London, 1888), p.87

Really complicated situations could arise only in games of chance in those early days, and these never attracted the attention of the rabbis. Nonetheless, it is remarkable that very early there was considerable interest in combinations, especially among the Jewish mathematicians. It seems that at least part of the impetus for the development of this topic stemmed from mysticism and astrology.

11.1 LETTERS AND WORDS

Of particular interest are the *Sefer Yetsirah* or *Book of Creation* and its commentaries.[2] Containing less than sixteen hundred words, much of the *Sefer Yetsirah* is devoted to enumerating various permutations of the letters of the alphabet. Critical opinion is divided as to when it was written. Gandz[3] dates it no later than the second century, but possibly much earlier. It is certainly not later than the eighth century. Among the major commentaries are those by Rabbi Saadya Gaon (892–942) and Rabbi Shabbetai Donnolo (913–970).

For the author of *Sefer Yetsirah*, the twenty-two letters of the Hebrew alphabet represent the building blocks of creation: '[God] drew them, hewed them, combined them, weighed them, interchanged them, and through them produced all of creation and everything that is destined to be created.'[4]

The number of combinations of the twenty-two letters taken two at a time[5] is given as follows: 'Twenty-two element-letters are set in a cycle in two hundred and thirty-one gates – and the cycle turns forwards and backwards ... How is that? ... Combine *A* with the others, the others with *A*, *B* with the others, and the others with *B*, until the cycle is completed.'[6]

2 On this book, see *Jewish Encyclopedia* XII, 602–3. References to *Sefer Yetsirah* are to chapter and paragraph.

3 S. Gandz, 'Saadya Gaon as a mathematician,' *American Academy of Jewish Research Texts and Studies*, vol. II (Saadia Anniversary Volume), 1943, 159

4 *Sefer Yetsirah* II 2

5 Boetius (sixth century) gave the rule for the number of combinations taken two at a time: $C_{n,2} = n(n-1)/2$. See G. Sarton, *Introduction to the History of Science* (Baltimore, 1927–48), vol. 1, 425.

6 *Sefer Yetsirah* II 4

The commentators explain that the description of one arrangement as 'forwards' and the other as 'backwards' refers to the fact that in 231 permutations the letters appear in the order of the alphabet, the earlier one in the alphabet followed by the later one, while in the 'backward' 231 permutations the order is reversed; also that the computation rule – '*A* with the others etc.,' means that with *A* in the first place there are twenty-one pairs and similarly for every one of the twenty-two letters of the alphabet – so that the total number of arrangements is $22 \times 21 = 462$.

A more general rule is: 'Two stones [letters] build two houses [words], three build six houses, four build twenty-four houses, five build one hundred and twenty houses, six build seven hundred and twenty houses, seven build five thousand and forty houses, henceforward go and calculate ...' [7]

To illustrate, since the longest word in Scripture consists of eleven letters, Rabbi Saadya actually computes the number of permutations of eleven letters, namely, 39,916,800.[8]

Donnolo gives the following proof for the rule that n letters can be arranged in $n!$ ways:

A single letter stands alone but does not form a word. Two form a word: the one preceding the other and vice versa – give two words, for twice one is two. Three letters form three times two that is six. Four letters form four times six that is twenty-four ... and in this way continue for more letters as far as you can count. The first letter of a two-letter word can be interchanged twice, and for each initial letter of a three-letter word the other letters can be interchanged to form two two-letter words – for each of three times. And all the arrangements there are of three-letter words

7 Ibid. IV 12. The number 5040, which represents the number of permutations of seven elements, seems to have had a special fascination for mystics. It is noteworthy that Plato (fourth century B.C.E.) specifies that the number of citizens in the ideal state shall be set at 5040. '... let the magistrate, whom we shall lay down as the greatest and most honourable consider what is proper to be done with the superabundance and deficiency of children and devise a method by which 5040 households alone may exist always,' *The Laws* V 10 (translated by George Burges, London, 1896, p.176). Plato explains that 5040 is chosen because 'there are continuous divisions of it from one up to ten.' Ibid. V 9 (Burges, p.171)

8 See Solomon Gandz, 'Saadya Gaon as a Mathematician,' p.160

correspond to each one of the four letters that can be placed first in a four-letter word: a three-letter word can be formed in six ways, and so for every initial letter of a four letter word there are six ways – altogether four times six making twenty-four words ... and so on. This is their number without vowel-points [the Hebrew letters are all consonants]; it cannot be increased nor diminished. And if there be found a wise man who will take all twenty-two letters together and re-arrange them and the vowels in this manner and according to this computation, and break them up in halves, thirds ... regardless of whether they are meaningful to him or not, he would have all the words of all the languages on earth. But the number is too great for flesh and blood to calculate ... Now whosoever believes without seeing it, let him believe, but he who does not believe let him see that it is so ...[9]

Then follows a tabulation of all the permutations of seven letters.

Sefer Yetsirah inspired generations of mystics to speculate on more complicated problems, such as determining the number of permutations of letters some of which are identical.[10]

11.2 CONJUNCTIONS OF PLANETS

From quite a different point of departure, namely, an investigation of the number of possible conjunctions of the various planets, Rabbi Abraham ibn Ezra (*d*.1167) was led to consider combinations of n things taken r at a time. He gave rules for expressing the number of combinations of n elements, taken r at a time in terms of combinations of lower order.

The subject is treated in a Hebrew work entitled *Ha-Olam* (The World). A Latin translation made in 1281 is extant.[11] Ibn Ezra

9 See N.L. Rabinovitch, 'Combinations and probability in Rabbinic literature,' *Biometrika* 57 (1970), 203–5

10 See Morris Turetzky, 'Permutations in the 16th century Cabala,' *The Mathematics Teacher* 16 (1923), 29–34

11 The Latin version appears in Marshall Clagett, *Nicole Oresme and the Medieval Geometry of Qualities and Motions* (University of Wisconsin Press, 1968), pp.445 ff. An English translation and commentary was given by Jekuthiel Ginsburg, 'Rabbi Ben Ezra on permutations and combinations,' *The Mathematics Teacher*, 15 (1922), pp.347–56. The passages quoted here are from Ginsburg's version.

prefaces his analysis with a theorem that the sum of the series of integers from 1 to n is $n(n/2 + 1/2)$.

Now we shall proceed to find the number of binary conjunctions – that is the combinations of two stars each. And it is known that there are seven planets. Now Jupiter has six conjunctions with the planets. [Eliminating Jupiter, pick Saturn, say, and it has 5 conjunctions with the remaining planets. Eliminating Saturn proceed as before, until we get a series 6 + 5 + 4 + 3 + 2+ 1.] Let us multiply then 6 by its half plus half of unity. The result is 21, and this is the number of the binary conjunctions.

We wish now to know how many ternary combinations are possible. We begin by putting Saturn and Jupiter, and with them one of the others. The number of the others is five. [Eliminate Saturn, and take Mars with Jupiter and with them one of the remaining four; then eliminate Mars and the empty place can be filled in 3 ways, and so on until we get a series of the ternary conjunctions containing Jupiter 5 + 4 + 3 + 2 + 1.] Multiply 5 by its half plus half of unity. The result is 15, and these are the conjunctions of Jupiter.

In the case of the conjunctions of Saturn we have four planets left [leaving Jupiter aside, and taking Mars, say, with Saturn. Then proceed as before getting a series 4 + 3 + 2 + 1]. Multiply 4 by half of itself plus one-half. The result is 10.

The conjunctions of Mars are ... 6.

The conjunctions of the Sun are ... 3, and the conjunctions of the [three] lower planets are 1. All together they are 35 and this is the number of ternary combinations.

For the quaternary conjunctions Rabbi Abraham ibn Ezra carries out the computation in the same manner and concludes: 'All together, there are 35 quaternary conjunctions.' Thereafter, the computation is omitted, and the results are set forth as follows:

We wish to find the quinary conjunctions. We find for Jupiter 15, for Saturn 5 and for Mars 1, together 21. The senary conjunctions are six for Jupiter and one for Saturn [i.e. excluding Jupiter], and there is one conjunction of all the seven planets. The total number is 120 conjunctions, all component [groups of] conjunctions are odd in number and they are divisible by 7.

11.3 A MAJOR BRANCH OF MATHEMATICS

The whole theory was perfected and advanced by the time of Rabbi Levi ben Gershon, who presents the topic of combinations and permutations as one of the major branches of arithmetic in his *Maasei Ḥoshev*, written in 1321.[12] The whole work is divided into two parts which we might describe as 'pure' and 'applied' mathematics respectively. In the first part, sixty-eight theorems are formulated in general terms and are followed by careful proofs. In the second or 'applied' part, detailed instructions are given for working numerical problems involving the theorems of the first or 'pure' part, but actual applications are not even mentioned. Six topics are covered: (1) addition and subtraction, (2) multiplication, (3) arithmetic and geometric series, (4) combinations and permutations, (5) division and extraction of square and cubic roots, and (6) proportions.

It would seem from this treatment that the theory of combinations had an important place in the mathematics of the time. Yet nowhere does Rabbi Levi ben Gershon indicate in what connection it might be found useful.

In the theoretical part of *Maasei Ḥoshev*, the section on combinations contains six theorems and, unlike the other topics, is introduced by a special preface which includes an explanation of the difference between arrangements in which the order of the elements is distinguished (permutations) and those in which order is immaterial (combinations). The terminology, which is apparently original with Rabbi Levi, though stylistically rather clumsy, is

12 The Hebrew text of *Maasei Ḥoshev* with a German translation and introduction and notes by Gerson Lange was published as *Sefer Maassei Choscheb – Die Praxis des Rechners, Ein hebraisch – arithmetisches Werk des Levi ben Gerschom aus dem Jahre 1321* (Frankfurt-am-Main, 1909).

Some passages are rendered into English in N.L. Rabinovitch, 'Rabbi Levi Ben Gershon and the origins of mathematical induction,' *Archive for History of Exact Sciences,* 6 (1970), pp.237–48

descriptive and precise. The proofs are remarkable for their use of mathematical induction, and are substantially the same as one finds in any modern text.

After deriving the general formula for the number of permutations of n things taken k at a time, Levi derives the general rule for the corresponding number of combinations, which he formulates in words in a manner which is equivalent in our notation to

$$\binom{n}{k} = \frac{n(n-1)\ldots(n-k+1)}{k!}.$$

The development is climaxed in Theorem 68 which reads:

If there be a given number of different elements of which are taken as many at a time as a second given number, yielding as many combinations as a third given number, and if the difference between the first given number and the second be called the fourth number, then taking as many at a time as the fourth number gives as many combinations as the third given number.

In modern notation this is simply

$$\binom{n}{k} = \binom{n}{n-k}.$$

For computation, in the second part of his book, Rabbi Levi ben Gershon illustrates the use of this formula in evaluating

$$\binom{8}{5} = \binom{8}{3} = 56.$$

11.4 AN APPLICATION TO PROBABILITY

Although, much of Rabbi Levi ben Gershon's work was known in Christian circles, there was apparently little or no familiarity with his theorems on combinations, and Marshall Clagett was unable to find 'any Latin author other than Oresme who demonstrates knowledge of these combinatorial rules.' As for Oresme, although he refers to such rules, he does not state them and moreover his own computation contains an error.[13] Nonetheless Nicole Oresme does utilize combinations in several contexts, which demonstrates that they were of practical use. Thus, he finds how many lines can be drawn from one to the other of four distinct points.[14] He also applies combinations to his *geometry of qualities*,[15] although this is likely an original application of his own adapted to this novel concept.

More importantly, apparently alone among his Latin confreres, Oresme applies combinations to probability. He takes 100 'ratios,' 2/1, 3/1, ... , 101/1, and by combining them two at a time, shows that there are 4950 possible different pairs which he calls 'ratios of ratios,' of which only 25 are of a type which he designates 'rational,' all the rest being irrational. The number 4950 is obtained by a rule which is equivalent to

$$\binom{100}{2} = \frac{100 \times 99}{2}.$$ [16]

He notes that for larger groups of whole-number ratios, for example, from 2/1 to 201/1, the relative frequency of 'rational ratios of ratios' is less and it decreases as ever larger groups are considered. From this he concludes that the question whether or not an unknown 'ratio of ratios' happens to be rational ought to be answered 'in the negative, since this seems more probable and likely ... since there are many more irrationals than others.'[17] He proceeds to infer that the same holds true for the ratio of any two unknown ratios, each of which describes a motion, time, or distance of one of the heavenly bodies, and thus it follows that

13 Marshall Clagett, *Nicole Oresme and the Medieval Geometry of Qualities and Motions: Tractatus de configurationibus qualitatum et motuum* (University of Wisconsin Press, 1968), pp.444–7

14 Nicole Oresme, *De proportionibus proportionum* and *Ad pauca respicientes*, edited and translated into English by Edward Grant (University of Wisconsin Press, 1966), p.257

15 Nicole Oresme, *Tractatus de configurationibus qualitatum et motuum*, ed. Marshall Clagett, p.205

16 *De proportionibus* ... , p.255

17 Ibid., p.251

When two motions of celestial bodies have been proposed, it is probable that they would be incommensurable, and most probable that any celestial motion would be incommensurable to the motion of any other sphere ...

[Consequently] In any instant it is necessary that celestial bodies be so related that in any moment there will be a configuration such that there never was a similar one before, nor will there be one after in all eternity.[18]

Thus, astrology which purports to predict the future on the basis of precise determinations of conjunctions and oppositions is rendered futile and false. Oresme justifies the implicit assumption that all individual ratios of ratios are equally probable by the argument that 'we have doubts as to which is necessary, so that we say of each that it is possible.'[19] This is just the Principle of Indifference.

Was Oresme influenced by his Jewish predecessors? Although a definite answer cannot now be given, two important facts must be mentioned which may have some bearing on this problem. First, Oresme quotes from Maimonides' *Guide of the Perplexed*, which was available in Latin translation in the first half of the thirteenth century.[20] If he knew the *Guide* well, the probabilistic concepts in it would not have eluded a mathematically acute mind such as Oresme's.

The second fact which may be relevant is that Rabbi Levi ben Gershon, towards the end of his life in 1343, wrote his tract *De harmonicis numeris* at the specific request of Philippe de Vitry, to whom Oresme dedicated his treatise *Algorismus proportionum* perhaps a decade later.[21] Levi is known to have had close relations with Christian clerics and at least one of them assisted him in his astronomical observations.[22] It may be that it was through Levi's

18 Ibid., pp.305–7. That an extension to infinity is unwarranted seems to have escaped Oresme's notice

19 Ibid., p.385. Both Grant and Clagett treat Oresme's probability concept as unique among the Latin schoolmen.

20 The citation is from the *Guide* II 36, and occurs in *De configurationibus* ... I 34, ed. Clagett, p.255.

21 See Marshall Clagett, *Nicole Oresme* ... , p.447; also Edward Grant's Introduction to *De proportionibus* ... , p.12

22 Bernard R. Goldstein, 'Preliminary remarks on Levi Ben Gerson's contribution to astronomy,' *Proc. Israel Acad. Sci. & Hum.* 3 (1969), 243

influence that the question of the commensurability of the celestial motions was introduced to the Parisian circle. Levi discusses the matter in his *magnum opus Milḥamot haShem*[23] and he regards it as very important in the light of his theory of knowledge.

11.5 COMBINATIONS IN GAMING

It seems that Tartaglia was the first to apply the theory of combinations to the cast of dice.[24] Even much later, it was only in the context of gambling problems that precise probability computations using combinatorial analysis were of interest. Applications to scientific problems were a long time in coming. Without the stimulus from these sources, the development of probability theory among the Jews could not move beyond the mathematical and philosophical fundamentals.

23 *Milḥamot haShem* v iii, 10. Grant, in his introduction to Oresme's *Ad pauca respicientes* (pp.111 ff.), marshalls some possible sources for the transmission of the problem to Oresme, but is unaware of the possibly key role that might have been played by Rabbi Levi ben Gershon. See also above p.140 n.10

24 See D.E. Smith, *History of Mathematics*, vol.2 (Dover, 1958), 526

12
The historical perspective

12.0 DEMONSTRATION AND DIALECTICS

It has been shown that a major element in the rabbinic ideology of probability is the principle 'Follow the Majority.' The development and application of this principle of statistical inference led to most of the ramifications of probability theory in Jewish literature which we have discussed. Now this is such a basic idea that it was no doubt recognized by men of every culture. In fact, it is not unlikely that primitive man, like a child, reasoned entirely from analogy, and 'Follow the Majority' represents the next higher stage of thought.

When the strict canons of deductive reasoning were formulated and clearly understood, attention was directed to the fundamental superiority of deduction over induction in yielding certainty. Aristotle (384–322 B.C.E.), who is often credited with inventing the syllogism, did not, however, overlook the importance of non-deductive inference. He distinguished between 'demonstrative' and 'dialectical' reasoning.

(a) It is a 'demonstration' when the premises from which the reasoning starts are true and primary, or are such that our knowledge of them has originally come through premises which are primary and true:

(b) reasoning, on the other hand, is 'dialectical,' if it reasons from opinions that are generally accepted. Things are 'true' and 'primary' which are believed on the strength not of anything else but of themselves; for in regard to the first principles of science it is improper to ask any further for the why and wherefore of them: each of the first principles should command belief in and by itself.[1]

The result of demonstrative reasoning is 'scientific knowledge' which is true and incontrovertible. But reasoning from 'opinion' is not certain. Thus Aristotle says: 'Now of the thinking stages by which we grasp truth, some are unfailingly true, others admit of error – opinion, for instance, and calculation, whereas scientific knowing and intuition are always true.'[2] Opinion derives from various sources, one of which is what Aristotle calls a 'probability.'

... a probability is a generally approved proposition; what men know to happen or not to happen, to be or not to be, for the most part thus and thus is a probability.[3]

What constitutes 'general' approval of a proposition? Those opinions are 'generally accepted' which are accepted by every one or by the majority or by the philosophers – i.e., by all or by the majority, or by the most notable and illustrious of them.[4]

It appears then that the limitations of the human mind are responsible for the fact that some propositions may not be known with certainty but only probably. The probable is that which is accepted by most men or by the wisest among them.

Premises that are probable can be used to form syllogisms, but the conclusions which they yield cannot be more certain than the premises.

Because, he looked for certainty, Aristotle gave priority to axioms and intuition. To this end he attributed to sense-perception

1 *Topica* I 1,100a27
2 *Analytica Posteriora* II 19,100b5
3 *Analytica Priora* II 27,70a3
4 *Topica* 100b21–23

the ability to discover the Universal in the particular.[5] Because in practice, though, there are few insights which can be claimed to be known with certainty, 'probabilities' and 'opinions' had to be admitted. Thus both demonstration and dialectics occupied Aristotle's attention and both were diligently studied by his followers and heirs in the medieval Arabic, Hebrew, and Latin traditions. Moreover, dialectic was regarded as the method most suitable for the law-courts. In Aristotle's words:

> If you have no witnesses on your side, you will argue that the judges must decide from what is probable; that this is meant by 'giving a verdict in accordance with one's honest opinion'; that probabilities cannot be bribed to mislead the court; and that probabilities are never convicted of perjury. If you have witnesses, and the other man has not, you will argue that probabilities cannot be put on their trial, and that we could do without the evidence of witnesses altogether if we need do no more than balance the pleas advanced on either side.[6]

The practical importance of dialectics notwithstanding, Aristotle does not examine carefully how probable judgments arise. His primary concern in his treatment of dialectics is with the deductive reasoning that is based upon probable premises. As for the premises themselves, his notion derives from the principle 'Follow the Majority.'

12.1 ETYMOLOGY AND MEANING

Yet it seems that Aristotle never moved beyond the stage of 'Follow the Majority.' Perhaps because he sought the universal in the particular, it was unacceptable to him to posit a proposition about a collective which is not also at the same time true about the individuals which constitute it. Thus his word *εἰκός* is translated

5 *Analytica Posteriora* II 19,100a16: 'for though the act of sense-perception is of the particular, its content is universal.'
6 *Rhetorica* I 15,1376a 16–23

into Latin as *verisimilis* namely, 'like truth' or plausible.[7] Other words which came to be used for the same concept, for example, Latin *probabilis*, French *vraisemblable*, German *wahrscheinlich*, all convey the same sense of similarity to or almost truth. While it was recognized that there might be exceptions to a 'probability' and to that extent 'opinion' may be liable to error, it was thought that truth and falsehood are two contraries, which, unlike other sets of contraries such as hot and cold, do not admit of mixing. In his discourse on rhetoric, Socrates argues that fraud results from verisimilitude 'When will there be more chance of deception – when the difference is large or small? ... and you will be less likely to be discovered in passing by degrees into the other extreme than when you go all at once.'[8] A little error is like a powerful poison contaminating the whole body of truth.

In the Middle Ages, the distinction between science and opinion was taken over and elaborated by a long line of Aristotelians. An example is Thomas Aquinas (1225?–1274), who points out that one's confidence in the truth of a proposition can be other than complete and so our opinions can be maintained in spite of an element of uncertainty, 'for opinion seems to have the meaning of something weak and uncertain.'[9] Nonetheless it is only insofar as a proposition recommends itself to us as being similar to truth that opinion inclines to it at all. Albertus Magnus inscribed at the head of his *Topicorum* the phrase *probabilia sunt verisimilia*.[10] It follows from this attitude that even if one can conceive of greater or lesser similarity to truth, it is utterly impossible to deal with the continuum of probability. Indeed, it seems likely that some logi-

7 For a textual analysis of the usuage of 'probability' in Aristotle and the early moderns beginning with Locke, see Siri Blom, 'Concerning a controversy on the meaning of "Probability," ' *Theoria* 21 (1955), 65–98

8 'Phaedrus,' *The Works of Plato*, vol. 2, translated into English by B. Jowett (Oxford, 1875), 139

9 Thomas Aquinas, *Exposition of the Posterior Analytics of Aristotle*, translated by Pierre Conway (Quebec, 1956), Lecture 44, sect.6, p.283

10 Quoted by A. Gardeil, 'La certitude probable,' *Revue des Sciences Philosophiques et Théologiques* 5 (1911), 248

cians recognized only the two categories, probable or 'for the most part' and improbable or 'for the lesser part,' without any distinction of degree.[11]

Terms like *aestimatio* and *existimatio* which were used as equivalent to *opinio,* although suggesting valuation, apparently did not imply a comparative scheme which could itself be made precise. Rather, as Aristotle had said, it was to 'indicate the truth roughly and in outline.'[12]

On the other hand, the Hebrew terminology is suggestive of a divergent conception. The most common Talmudic word for probability is *'omed* (אומד) with its derivative forms. This is used also in the sense of 'appraise' as, for example, in appraising the value of an object, or in the sense of the modern English 'estimate' as in estimating the height of a hill. The root is *md* (מד) – to measure.[13] Another word that is used in the Talmud is *sevara* (סברא), meaning reasonable or rational. It is appropriate in some instances where *'omed* might not be used, such as in evaluating an opinion for its reasonableness when a numerical measure is not applicable. The root is *br* (בר) – to be bright or clear.[14] In modern Hebrew the technical word for mathematical probability is derived from *sevara*,[15] which is somewhat surprising since *'omed* may be historically as well as semantically more appropriate.

11 A. Gardeil in 'La certitude probable,' p.260, describes the view of Aquinas and other medieval Christian thinkers as follows: 'L'élement formel du probable est la ressemblance ou approximation du vrai ... qualité par laquelle sont exclues de la probabilité les propositions douteuses, peu probables ou moins probables.' Again on p.264, 'il n'y a pas de moins probable.'

See, however, Edmund F. Byrne, *Probability and Opinion* (The Hague, 1968), p.224, who disputes Gardeil's conclusions

12 *Ethica* I 3:1094b20

13 See Marcus Jastrow, *A Dictionary of the Targumin, the Talmud ...* , vol. 1 (New York, 1950), 75. Perhaps it is significant that already in the Biblical example (Job 21:17) of probability reasoning (see above p.26), Job refers not merely to a majority or minority: rather he uses a quantitative term, 'How often is it ... ,' literally *How many* (כמה).

14 Marcus Jastrow, *Dictionary* ... , vol. 2, 951

15 It is perhaps worth mentioning that in the thirteenth-century Latin version of the *Guide* (II 22) in describing the tentative character of Aristotle's theories of the supralunar spheres the word used is *verisimilia*, corresponding to the Hebrew *sevara*

In the Hebrew of the Middle Ages, apparently under the influence of Arabic, a pair of contraries was introduced to denote 'probable' and 'improbable,' namely *karov* (near) and *raḥok* (remote). The same usage appears in English when we say 'the chances are remote that ... ' The adverbial as well as the related forms appear with modifiers indicating degree. These range from 'exceedingly near'[16] to 'near'[17] through 'a little remote'[18] to 'very very remote.'[19] Also common is the combination of the adjectives 'near' or 'remote' with *efshar* (contingency or possibility). Thus Rabbi Samuel ibn Tibbon renders Maimonides: 'It happens many times for it is a *near contingency*';[20] and at the other end of the scale – 'a remote contingency.'[21]

Occasionally, especially when dealing with high probabilities, more explicit phrases occur. Thus, for example, the following seem to be almost equivalent usages: 'near certainty,'[22] 'near reality,'[23] 'near demonstration,'[24] and 'near complete proof.'[25]

12.2 EQUIPROBABILITY

Perhaps the crucial concept is that of equiprobability. Probabilities derived from majorities can be conceived of as approximations to truth, while the attendant uncertainty reflects either the in-

16 Maimonides, *Mishneh Torah, Sanctification of the New Moon*, 18:7, '... exceedingly near, it happens frequently ...'

17 Samuel ibn Tibbon's Hebrew version of Maimonides' *Guide* III 50

18 Rabbi Abraham ibn Ezra, *Commentary on Isaiah*, 1:6

19 Ibn Tibbon's Hebrew version of the *Guide* II 19

20 *Commentary on the Mishnah, Bekhorot* I 4

21 *Guide* I 32

22 'קרוב לוודאי,' Rashi to *Betzah* 8b s.v. מותר and s.v. וחד where the phrase is as it were defined: שכיחא וקרובה לוודאי – 'frequent and near certainty'; Tosafot *Pesaḥim* 9a s.v. ואם; Rabbi Nissim ben Reuven, *Commentary to Alfasi, Kiddushin* 39a s.v. א"ר אסי

23 'קרוב מהמציאות,' Maimonides, *Sefer HaMitzvot*, Negative Commandments, no. 290. See above p.111 where the context is quoted. However, it seemed best to translate it there – 'almost a certainty.'

24 'קרובות למופת,' Maimonides, *Guide* II 19. Pines translates 'near demonstrative certainty' (p.309).

25 'קרוב לראיה גמורה.' Rabbi Joseph Colon, *Responsa*, no. 129

adequacy of our knowledge or even an aberration in nature itself. Thus, Aristotle ascribed unfailing regularity to the heavenly movements, whereas in the sublunar world 'mistakes are possible in the operations of nature also ... and monstrosities will be failures in the purposive effort.'[26] Certainty is precluded because perfect knowledge is denied man inasmuch as earthly existence is imperfect. There are hindrances which are due to man's weakness and there are those which originate because of some lapse in the world. 'In natural products the sequence is invariable, if there is no impediment.'[27] Now knowledge of the 'mistakes' of nature is a contradiction in terms: it is not possible to *know* a mistake. To the extent that there is knowledge, it refers to the majority or usual run of things and it is therefore as near to truth as one can come.

However, this conception is totally unsuitable for equally probable alternatives. Thus Rabbi Levi ben Gershon argues that the term 'knowledge' applies only to that which makes definite 'which of the various contingencies will come to pass.' Without this specification

> ... we do not call it knowledge; rather it is perplexity and confusion. For then we say that we are perplexed and unable to estimate whether this alternative or this or that will come to be. The more numerous are the alternative contingencies the greater is the perplexity. Now perplexity and confusion are contrary to knowledge, and certainly so in this manner. For any thing that comes into being has preceding its existence contrary contingencies, and in a thousand years, say, these must be almost infinite in number. This is so because any one of the intermediate events can either be or not be, and if it be not, it is possible that such and such be the case or not. Now if you continue along these lines it will be established that the final event which comes to pass as a result of all the preceding intermediaries having come to be, will be one of an endless number of contraries. This, it seems to us, is complete perplexity and confusion, for one

26 *Physica* 199b1

27 *Physica* 199b25. See above p.26 for a discussion of Job 21: 17–18, where a similar idea is expressed

must be perplexed as to whether it will be thus or thus or thus in an infinite number of ways. Now perplexity and doubt are less when the possible contraries are fewer, and so an answer which specifies a thing more is more informative.[28]

As long as it is the knowledge of particular things that is sought, the least that will pass for knowledge is a verisimilitude and that requires nothing less than 'Follow the Majority.' When one foregoes the demand for knowledge and seeks instead rational grounds for acting in the absence of knowledge of things in themselves, it may be useful to have information about the number of possible alternatives or about a set of things as a whole rather than about each individual or even about a majority of the elements in the set. That a minority yields a probability in the same sense as a majority, with the difference between them only a matter of degree not of kind, is the fundamental idea without which modern probability conceptions could not develop. Apparently such an idea can take root easily only when considering equal probabilities which must therefore all be less than majorities.

Whether or not a consistent interpretation of equiprobabilities is formulated, as soon as they are clearly recognized as probabilities and used in computations, the idea that probability is a verisimilitude can no longer be sustained. Thus the way is opened for consideration of a metric of probability.

It is difficult to exaggerate the magnitude of the obstacles in the way of accepting equiprobabilities as being meaningful in their own right. Majorities were explained by laws of nature: minorities were seen as due to chance. In dealing with a number of equally probable alternatives, they must all be ascribed to chance. Can there be a rational measure of chance? There was a strong philosophical bias against the formulation of any kind of 'laws of chance.' Aristotle had argued the case cogently and compellingly and his conclusion is unambiguous. '... chance is a thing contrary to rule ... For "rule" applies to what is always true or true for the most part,

28 *Milḥamot haShem* III 2

whereas chance belongs to a third type of event. Hence, to conclude, since causes of this kind are indefinite, chance too is indefinite.'[29]

Although one might suppose that gradations of probability arising from sums of equal probabilities would naturally suggest themselves when dealing with numbered mixtures of known composition, it is possible that these did not concern the Greeks. For them, only unenumerated majorities were of importance. Thus Aristotle speaks of the following as examples of probabilities: 'the envious hate,' 'the beloved show affection.'[30] These seem remarkably like the *ḥazakah* presumptions we have met in the Talmud.[31]

12.3 GREEK COMBINATORIAL ANALYSIS

Games of chance were avidly pursued in Greece, yet it seems that the equality of alternatives was neither understood nor sought. The matter was studied carefully by Sambursky,[32] who found that in games with the *astragalus* which is of asymmetric shape 'no account was apparently taken, in fixing the values of the faces, of the relative frequencies with which they turn up.' He points out, too, that the *Venus* throw (1, 3, 4, 6) for four *astragali* thrown together was valued higher than all the other less frequent throws, including the throw (6, 6, 6, 6) the probability of which is much lower.

Moreover, Sambursky cites reports of computations of combinations and permutations of the letters of the alphabet which lead him to the judgment that 'the combinatorial analysis of the Greeks was faulty.' A figure is given for Xenokrates of Chalcedon (fourth century B.C.E.) who supposedly calculated that the number of syllables that can be written with all the letters of the alphabet is 1000

29 *Physica* II 5, 197a17–20; See also *Metaphysica* VI 2–3 and XI 8. It may be significant that the Talmudic ideas on probability were developed before Jewish thought was exposed to Aristotle in the eigth and ninth centuries.

30 *Analytica Priora* II 27, 70a3

31 See above p.114

32 S. Sambursky, 'Possible and probable in ancient Greece,' *Osiris* 12 (1956), 35–48

$\times 10^9$. Sambursky notes: 'As there are few syllables with more than four letters the figure given seems to be a gross exaggeration.'[33] Unfortunately, there is little information available and no details on how these computations were made.

It is not till Boethius in the sixth century that we find the explicit rule for computing the number of combinations of n things taken two at a time,[34] the selfsame rule which is given in *Sefer Yetsirah.*[35] Furthermore, it is not connected with any probability consideration.

12.4 INHERITANCE PROBLEMS: MATHEMATICAL EXPECTATION

It would be surprising if there was to be found much familiarity with Talmudic probability in non-Jewish sources. Although Jewish writings exerted considerable influence on both Arabic and Latin scholars, it was mostly limited to philosophical, scientific, and theological works, not legal texts. These were not translated or generally taught outside the Jewish fold. Even the exceptional non-Jewish scholar who acquainted himself with the Talmud sought in it only theological matters. Nonetheless, there were some contacts, especially with Moslem law, mainly through personal interchanges. One legal subject which had far-reaching mathematical ramifications in both Moslem and Jewish jurisprudence is inheritance.[36] It is important, also, as a field of application, for probability.

We have already seen that, when there are several litigants or heirs whose claims are equal, the Talmud assigns them equal shares.[37] In Arabic manuals of arithmetic, legacy problems loom

33 Ibid., 44. See also D.E. Smith, *History of Mathematics* vol. 2 (New York, 1958), 524, who reaches the same conclusions

34 D.E. Smith, *History of Mathematics*, vol. 2, 52

35 See above p.143

36 See Solomon Gandz, 'The algebra of inheritance,' *Osiris* 5 (1938), 319–91 for an Arabic treatise; 'Saadya Gaon as a mathematician,' *Amer. Acad. Jew. Research Texts and Studies* II (1943), 166–81, where a Jewish text is discussed

37 Above p.76

large. It seems that at least in oral discussions between Jewish and Moslem sages, the different approaches of the Talmud and Islam were amplified. This appears from a remark of Rabbi Abraham ibn Ezra in an analysis of just such as inheritance problem in his *Sefer haMispar*.[38] Because this may be an important link in establishing a chain leading right up to the beginnings of modern probability theory, it is worth quoting in large part.

Question: Jacob died and his son Reuben produced a deed duly witnessed that Jacob willed to him his entire estate on his death. The son Simeon also produced a deed that his father willed to him half of the estate. Levi produced a deed giving him one third, and Judah brought forth a deed giving him one quarter. All of them [the deeds] bear the same date. The gentile sages would divide the estate in accordance with the ratio of the face value of each, while the Jewish sages divide it in proportion to each one's claim.

Thus the mathematicians [i.e. the gentile sages] say that the amount is 1 and when you add to it 1/2 plus 1/3 plus 1/4 the sum is 2 1/12 ... In short, Simeon takes half of Reuben's share, and Levi one third of Reuben's share and Judah one fourth of Reuben's share.

Ibn Ezra then illustrates by taking an example where the estate is worth 120 and he calculates each share, Reuben receiving 57 3/5 and so on. Then he proceeds to explain the Talmudic rule.

In accordance with the view of the Jewish sages, the three older brothers say to Judah, 'Your claim is only on 30 [one fourth], but all of us have an equal claim on them. Therefore, take 7 1/4 which is one quarter and depart.' Each one of the brothers takes a similar amount. Then Reuben says to Levi, 'Your claim is only on 40 [one third]. You have already received your share of the 30 which all four of us claimed: therefore take one third of the [remaining] 10 and go.' Thus Levi's share is 10 5/6 [that is, $30 \times 1/4 + 10 \times 1/3$] ... Reuben also says to Simeon, 'Your claim is for only half of the estate which is 60, while the remaining half is all mine. Now you have already received your share of the 40, so that the amount

38 Moritz Silverberg, ed. (Frankfurt-am-Main, 1895), p.57–8

at issue between us is 20 – take half of that and depart.' Thus Simeon's share is 20 5/6 [i.e. $30 \times 1/4 + 10 \times 1/3 + 20 \times 1/2$] and Reuben's share is 80 5/6 [$= 30 \times 1/4 + 10 \times 1/3 + 20 \times 1/2 + 60 \times 1$].

From the casual way in which Ibn Ezra refers to the divergent views of the gentile and the Jewish sages, identifying the former with the mathematicians, it would seem that this type of problem was commonly discussed and both methods of attack were well known. If so, it is not implausible that even among the 'mathematicians' there were some who were at least familiar with the Talmudic probability idea underlying the Jewish solution.

It is worth noting that we have here an instance of the important concept of 'mathematical expectation' which in its modern form is apparently due to Leibniz,[39] although the idea already appears in Pascal.[40] The 'mathematical expectation' is defined as the product of the possible gain by the probability of attaining it.

In a sense it may be thought of as the benefit represented by a probability. The underlying idea is that one can compare a large probability of a small gain to a small probability of a large gain. If a man has a choice of paying ten cents for (a) a ticket giving him one chance out of ten of winning one dollar or (b) one ticket out of a hundred for a purse of $20, is there a best way to choose? The mathematical expectations for choices (a) and (b) are,

$$1/10 \times 1 = 1/10 \text{ and } 1/100 \times 20 = 2/10,$$

respectively. Consequently, choice (b) is to be preferred. Thus the concept of expectation yields a decision rule.[41] The mathematical expectation is closely related to the commercial concept of 'futures,' which is met with already in the Mishnah.

This occurs in a discussion of the present value of the marriage settlement. Jewish law provides that a woman is to receive a fixed

39 See J.M. Keynes, *A Treatise on Probability,* 3rd ed., p.311, n.1

40 *Pensées*, in *Œuvres de Blaise Pascal*, vol. IV, ed. Henri Massis (Paris, 1927), 43–4: 'Pesons le gain et la perte ... l'incertitude de gagner est proportionnée à la certitude de ce qu'on hasarde, selon la proportion des hasards de gain et de perte.'

41 See Harold Jeffreys, *Theory of Probability* (Oxford, 1939), p.31

sum from her husband's estate on termination of the marriage by death or divorce. This represents a prior lien on the husband's property. The Mishnah states the problem: 'One estimates how much would a man pay [now] for this woman's marriage settlement [which will be due her] if she will be widowed or divorced, but if she dies [first] her husband will inherit her [and the purchaser will get nothing].'[42]

In the Gemara,[43] it is pointed out that in general there is a difference in value between the husband's right and that of the wife, apparently based upon the life-expectancy differential. This is explicitly stated by Maimonides in another connection where he remarks about women, 'their lives are generally shorter than those of men.'[44] In a gloss of the Tosafot a possible reason is suggested: 'perhaps because a woman is often endangered in childbirth.'[45]

Maimonides observes that, in addition to factors such as health and age which affect the probabilities, one must also take into consideration the variable utility of money. 'The present value of a large marriage settlement is not [exactly] proportionate to that of a small marriage settlement. For a marriage settlement of 1000 zuz that can be sold at a present value of 100, if the face value were 100 it could not be sold for 10 but rather for less.'[46]

Bayes[47] assumed that a 1/10 chance of getting $1 is as valuable as a certainty of getting 10 cents. Maimonides apparently felt that if one is to take a risk at all, it is worth relatively higher stakes if the winnings will be larger. It is now universally conceded that expectations of gain may not be assumed to be additive because of various psychological factors, but there is no unanimity on what constitutes an acceptable measure of utility.[48]

42 *M. Makkot* I 1

43 *Makkot* 3a

44 *Commentary on the Mishnah: Niddah* v 6

45 *Ketubot* 83b s.v. מיתה

46 *Mishneh Torah: Edut* XXI 1

47 Thomas Bayes, 'An essay towards solving a problem in the doctrine of chances,' *Philosophical Transactions*, vol.53, 376–98

48 For an interesting attempt to develop a logic of preference, see Nicholas Rescher, 'Semantic foundations for the logic of preference,' in *The Logic of Decision and Action*, ed. Nicholas Rescher (University of Pittsburgh Press, 1966). pp.37–79

12.5 LEGACIES, WAGERS, AND PROBABILITY

It is well known that 'inheritance' and related 'partnership' and 'wager' problems were very popular and they are frequently found in Arabic manuscripts. The great ninth-century mathematician Al-Khuwarizmi who, it is said, gave his name to algebra, devoted about half of his treatise, to problems on legacies.[49] From there they found their way into Latin and other European languages. Ore[50] has found a number of such problems in medieval Latin and Italian sources, including the celebrated ones of Chevalier de Méré which Pascal solved in 1654, thus initiating the birth of modern mathematics of probability.

The so-called division problem was the subject of the correspondence between Pascal and Fermat in which it has been claimed they laid the foundations for mathematical probability. Essentially it concerns two players of equal skill who leave the table before the game is completed. The stakes, the score necessary to win, and the actual scores of the players until quitting time are known. It is required to divide the stakes. This is done in accordance with each player's chances of winning if the game were to go on to the end. The problem has been found in the printed works of Tartaglia (1556) and Cardan (1539) and earlier still in Luca Pacioli's *Suma* which appeared in 1494.[51]

Ore reports it in Italian manuscripts as early as 1380, and connects it with the distribution and inheritance problems of the Arabs. Indeed, he conjectures, 'It appears likely that also in Pascal's circle in the mathematical academy, the simplest probability considerations were known.'

It is tantalizing to speculate on the possibility that somewhere in those collections of problems and solutions that were read by men like Tartaglia and Cardan early in the sixteenth century, there

49 This work introduced algebra to Christian Europe when it was translated by Robert of Chester and then again by Gerhard of Cremona. The chapters on inheritance with translation and commentary appear in Solomon Gandz, 'The algebra of inheritance,' *Osiris* 5 (1938), 319–91.

50 Oystein Ore, 'Pascal and the invention of probability theory,' *Amer. Math. Monthly* 67 (1960), 409–19

51 D.E. Smith, *History of Mathematics*, vol.2, 529

might have been an exposition of views of 'the Jewish Sages.' That these were known to some Arab savants seems beyond doubt, for Jewish influences on Islamic law are well established. Gandz surmises that 'The Muslim law on the legacy in his last illness seems to have been framed with the intention of contradicting the Talmudic law.'[52] While he is, of course, referring to general principles, it does not appear unlikely that the same opposition extended to details including those aspects involving probability.

12.6 FOURTEENTH-CENTURY PROBABILISM

'Since the fourteenth century the stream of sceptical empiricism has flowed strongly in European philosophy.'[53] The initiators of this movement began by meditating on the relation of faith to philosophy and the attempts at mediating the confrontation between religion and Aristotle. Their efforts were deeply influenced by and to a considerable extent modelled on the prior venture by Maimonides to synthesize the contents of revelation and reason.[54] Maimonides' great work *The Guide of the Perplexed* was translated into Latin early in the thirteenth century and was widely studied in Christian circles.

Basic to Maimonides' position is the view that natural philosophy offers probable theories rather than such as are necessarily true to explain the major questions of concern.[55] This doctrine of 'probabilism' found exponents in various schools of Latin scholasticism. Although in its Latin version it apparently remained rooted in the Aristotelian conception of the 'probable,' its major thrust was to undermine Aristotle's insistence on the certainty that attends 'scientific' knowledge, thus greatly expanding the relative scope of that part of knowledge which is only probable. This opened the way for a reconsideration of the meaning of probability as well as for a new view of natural science itself.

52 S. Gandz, 'The algebra of inheritance,' p.332, n.38
53 A.C. Crombie, *Augustine to Galileo*, 2nd ed., vol.2, (London: Mercury Books, 1961), 35
54 See Etienne Gilson, *Le Thomisme*, especially pp. 27, 47, 59, 60, 111
55 See above pp.123, 137

Furthermore, in the characterization of 'probable' propositions, even as restricted to the framework of dialectics and disputation, refinements introduced by Maimonides were adopted. Thus, instead of Aristotle's requirement of acceptance by most men or by the philosophers,[56] Maimonides maintains that the evidence for the probability of a proposition can be assessed only by an unbiased and morally stable seeker of truth trained in the intellectual disciplines.[57] While it may be difficult in practice to find a suitable impartial judge, it seems much more reasonable to expect that a man endowed with the requisite qualities will arrive at a true evaluation of probability than that this will be accomplished by a majority vote. This Maimonidean criterion is given, for example, by Nicholas of Autrecourt, one of the most radical of the fourteenth-century probabilists, who writes that, where there are arguments supporting each of two opposing conclusions, the correct comparison of the respective degrees of probability can be made 'by a lover of truth ... who is not moved to one or the other side.'[58]

56 *Topica* 100b21–23. The passage is quoted above, p.153

57 *Guide* II 23 (Pines, p.321). The passage is quoted above, p.139. See also *Mishneh Torah: Sanhedrin* XXXIV 2. However, while the *Guide* was well known to the schoolmen, it is unlikely that many knew the *Mishneh Torah*, except perhaps by hearsay.

58 *Exigit Ordo Executionis*, ed. J. Reginald O'Donnell, *Medieval Studies* 1 (1939), 203. Nicholas remarks that not all arguments are conclusive and for a proposition where no demonstrative proof is given, there may be evidence supporting both the proposition and its negation, and the relative strength of these evidences needs to be evaluated.

'Et si istae rationes non reperirentur omnino concludere, tamen probabilis est positio et probabilior rationibus conclusionis oppositae. Si enim habeant rationes qui tenent conclusiones oppositas, dicant eas et faciant super his comparationem amatores veritatis et credo quod cuilibet non magis affectato ad unam partem quam ad aliam apparebit gradus probabilitatis excedens in his rationibus.'

J. Reginald O'Donnell points to Maimonides as a possible source of Nicholas's atomism which 'was a current doctrine at Paris in the 13th and 14th centuries' ('The philosophy of Nicholas of Autrecourt and his appraisal of Aristotle,' *Medieval Studies* 4 (1942), 108, n.73). The case for Nicholas's dependence on Maimonides for that doctrine is developed by Julius R. Wein-

Only a few major studies of probability concepts in the Latin scholastics are known to me, and these each deal mainly with one or at most a few authors.[59] Until the field is thoroughly investigated, obviously a comparative analysis cannot be made. As far as seems to be known at present, a form closest to 'mathematical probability' was reached in Oresme's speculations, of whom Clagett says: 'the more that one examines closely the works of Oresme, the surer he becomes that Oresme was affected by the probabilistic and skeptical currents that swept through various phases of natural philosophy in the fourteenth century.'[60]

In the case of Oresme, as we have seen,[61] there may also be some dependence on Rabbi Levi ben Gershon who, although considerably older than Oresme, certainly had contact with the Parisian school. Rabbi Levi applied probabilistic concepts even to

berg in his study *Nicolaus of Autrecourt* (Princeton, 1948) pp.84 ff. The parallelism in such a fine detail as the characterization of the probable would seem to be additional evidence of Maimonides' influence on Nicholas.

Weinberg cites other Latin medievals who dispensed with Aristotle's 'idea that the probability of a proposition can be determined by a census' (ibid., pp.120 ff.), although strangely he does not make the connection with Maimonides.

59 I have already indicated (above, p.156 n.11) the nature of the conclusions reached by A. Gardeil in his papers of 1911 in which he studied dialectical probability in some of the major scholastics.

Recently, Edmund F. Byrne did a very thorough and apparently exhaustive examination of probability in Thomas Aquinas (*Probability and Opinion*). The work is limited to Aquinas only and, in any case, what emerges from this study is that there was no real progress in interpretation of probability other than within the confines of Aristotelian dialectics (see above, p.156 n.11).

Weinberg (*Nicolaus of Autrecourt*) summarizes the views of a number of fourteenth-century writers on probability as conceived of in disputation. The most radical position he mentions is that of Nicholas of Autrecourt which I have referred to briefly in the text.

Finally Edward Grant and Marshall Clagett in their editions of Oresme's work (already referred to) emphasize the novel nature of Oresme's ideas on probability, but they do not relate them to any contemporary or earlier work.

60 Marshall Clagett, *Nicole Oresme* ... , p.12. See above p.150 for a reference to Oresme's quotation from Maimonides.

61 See above, p.150

the distribution in the population of different age levels. In a comment on Moses' census of Israel, he notes: 'You will find it is usual that the people close to age twenty are very much more numerous than the people close to age sixty, for those that reach that number [of years] are few in number.'

He remarks further that for it to be otherwise would indicate 'powerful exercise of Divine providence to keep them alive for the possible term.'[62]

The recognition that there is a distinction between the 'possible' and 'probable' life-span is only the first step, but nonetheless an important one, to the development of actuarial theory.

As noted above, with respect to the possibility of the incommensurability of the celestial movements, it may very well be that it was Rabbi Levi who introduced that question in his milieu. The role of oral transmission cannot be overrated in matters of this sort. Probabilistic concepts, too, may have arisen in discussions between Rabbi Levi and the Christian scholars who were his friends. In any case, the influence of a man of Rabbi Levi's stature must have been very great, since he had access in the Hebrew to much learning from Jewish and Arab sources which still was not available in Latin, and he lived at a time and in a place where his knowledge was held in high regard by Christian clerics and the Jewish community enjoyed relative peace. This combination of circumstances, so unusual in Western Europe in the Middle Ages undoubtedly was of greater significance than appears from what has been gleaned so far from the available written records, especially since even the extant relevant manuscripts have not yet been thoroughly studied.

62 *Commentary* to Numbers 1:1. See also his commentary to Deuteronomy 4:3 and תועלת הי"ג

13
Conclusions

13.0 WHAT IS 'RATIONAL'?

Jehoshaphat, King of Judah, is praised in the Bible as one who 'did what was right in the sight of the Lord.'[1] His outstanding accomplishment was the improvement of the administration of justice. His charge to the judges was, 'Consider what you do, for you judge not for man but for the Lord; he is with you in giving judgment.'[2] The rabbis explain that a man might be overawed by the responsibility to judge for God and, fearing that truth cannot be discovered, withdraw entirely from the judicial task; therefore he is told: '*He is with you in judgment* – you have only [to be concerned with] what your eyes behold.'[3] A medieval commentator adds: 'It is the same for matters that depend upon inference, the judge has only [to be concerned with] what his heart [mind] sees.'[4]

The entire rabbinic literature is dominated by one theme. It is all an elaboration of the creative impulses generated by the tension

1 II Chronicles 20:32
2 Ibid., 19:6
3 *T. Sanhedrin* I 4
4 Rabbi Shlomo ben Meir, *Commentary on Bava Batra* 131a, s.v. ואל. See also the reference to this statement in the passage quoted from Rabbi Asher ben Yeḥiel, above p.128

between the two divine imperatives: (a) 'to do justice'[5] which summons man to action and, (b) 'You shall not pervert justice'[6] which freezes man into inaction for lack of the kind of certainty that is beyond man's ken. So that man will not shrink from his God-given duty, the assurance is forthcoming that *'He is with you in judgment'* if only one 'tries to bring the matter to justice and truth.'[7] The judicial process demands the utilization of every mode of reasoning and rational argumentation which is capable of honestly motivating sincere conviction with respect to objective truth. In fact, in the real world, there is no knowledge other than probable knowledge. The judge has 'only what his eyes behold,' in a figurative sense, but even empirical observations – literally 'what his eyes behold' – are subject to error of diverse kinds.[8] The confidence that one has in 'probable' knowledge rests ultimately upon one's trust in the human intellect which is God's gift to man: 'He is with you in judgment.' The rational faculty enables one to weigh the evidence and provided one makes every effort to avoid error and to steer clear of injustice, one may follow what appears to be reasonable and convincing evidence, subject to suitable safeguards depending upon the particular circumstances and subject, too, to such regulations as are prescribed in the Law.

Thus we have seen that the rabbis conceived of probability in different ways. In some situations only a comparative scheme was defined; in others a metric of probability was formulated. Quantitative probability was not understood in the same way in every application. A frequentist view is more appropriate for mixtures, while elsewhere a range of logical alternatives is considered. Never is there a suggestion that one particular interpretation is exclusive, although there are, as we have seen, discussions as to which interpretation is appropriate in a given instance. The same author felt free to use one or another view of probability as the occasion demanded. The attitude seems to have been that every interpreta-

5 Micah 6:8
6 Deuteronomy 16:19
7 Rashi to *Sanhedrin* 6b, s.v. אלא on the passage quoted above from the *Tosefta*
8 See *Bekhorot* 18a: 'It is not possible to make precise [measurements].'

tion that makes sense in some application is legitimate in that setting and all admissible interpretations complement each other, for they are all partial approaches to attain the kind of conclusions which warrant making decisions.

It is not surprising that starting out from a position in which jurisprudence is regarded as the model for rational argument, a leading logician of our time develops an outlook similar to the rabbinic one. Stephen Toulmin writes:

> To say that the term 'probability' cannot be analyzed in terms of frequencies or proportions of alternatives ... shows that they are to be regarded ... as different types of grounds, either of which can properly be appealed to, in appropriate contexts and circumstances, as backing for a claim that something is probable or has a probability of this or that magnitude.[9]

13.1 THE ANTIQUITY OF PROBABILITY AND DECISION THEORY

The materials analysed in the preceding chapters demonstrate that probabilistic reasoning has a very long history, much longer than has heretofore been thought. It is not the characteristic creation of recent centuries which has by virtue of its modernity rapidly become the major mode of rational enquiry; rather it has its roots in very early times and, at least in the Jewish tradition, was always a principal factor in human affairs.

Not only is the general concept of probability an ancient one: the axioms of the probability calculus as well as almost all of the various interpretations which are seriously entertained today have their origins in the early Middle Ages and beyond. Even some of the enigmas which beset modern philosophers of probability were puzzled over already many generations ago.

The awareness of the historical antecedents enhances the attractiveness of the view that the several conceptions of probability all have a place in the arsenal of rational thought. By refusing

9 Stephen E. Toulmin, *The Uses of Argument* (Cambridge, 1958), p.69

to admit some of them, significant areas of human endeavour would be deprived of decision procedures. In the last analysis, it is out of the need to provide ample grounds for decisions that probabilistic, and perhaps all reasoning, derives its impetus, and it is an indication of the validity of human reasoning that it successfully meets that need. 'By wisdom a house is built, and by understanding it is established.'[10]

תושלב"ע

10 Proverbs 24:3

APPENDIXES

Appendix A

GLOSSARY OF HEBREW TERMS

Amora *title of the teachers who taught during the third to the fifth centuries and whose discussions are recorded in the Talmud*
av *father*
ben *son*
binyan av *a rule of rabbinic interpretation equivalent to an analogy*
'efshar *contingent, possible*
Gemara *that part of the Talmud other than the Mishnah, consisting of the teachings of the Amoraim*
goral *lot*
ḥallah *offering from the dough*
ḥazakah *a presumption*
ḥelesh *a kind of lot*
ḥevel *portion or lot*
kab *a dry measure*
kal va-ḥomer *a rabbinic rule of inference; a kind of analogy*
karov *near; probable*
'omed *estimate of probability; conjecture*
orlah *the fruit of a tree during its first three years is forbidden and is called* orlah
payis *a kind of lottery for assigning the daily tasks to the priests in the Temple*

pur *lot*
rabbi *master, an honorific title accorded to teachers of the Law*
raḥok *remote; improbable*
Sanhedrin *the supreme judicial and legislative authority in ancient Israel*
seah *a dry measure*
sevara *opinion; reasonable argument*
shekel *a monetary unit*
Talmud (1) *learning*; (2) *the comprehensive designation for the Mishnah and the discussions on it in the academies up to the end of the fifth century*
Tanna *title of the scholars during the first two centuries* C.E., *whose views are preserved in the Mishnah, Tosefta*, etc.
tefillin *ritual objects worn on the head and arm containing Biblical passages setting forth the fundamentals of the faith*
teku *'let it stand,' a technical term to describe an undecidable proposition*
terumah *the offering to the priests from the crops. It is sacred and may be eaten only by priests under a strict discipline of ritual purity*
Torah (1) *teaching*; (2) *the name applied to the Five Books of Moses*; (3) *a comprehensive designation of the entire sacred Jewish literature*

Appendix B

GLOSSARY OF MATHEMATICAL AND TECHNICAL TERMS

This list includes only terms undefined in the text or such as are used in several places without being defined in each.

Acceptance Rule *a rule specifying the conditions under which* (1) *an hypothesis is accepted*; (2) *a set of products is approved*

Conditional Probability *the probability of an event E, given that another event* E_0 *has already happened; more generally, the probability of a set T, given a restriction to the set M*

Decision Rule *a rule specifying which procedure to follow, corresponding to the various possibilities for the observed data*

Independent Probabilities *two events M and T are said to be independent if the conditional probability of each one, given the other, is the same as its probability without this restriction*

Law of Large Numbers *if p is the probability of an event, then the relative frequency of the occurrences of that event in a number of trials will tend to p as the number of trials increases to infinity*

modulo *see residue*

operating characteristic *a function that gives for every value of the proportion of defectives in a lot the corresponding probability of accepting that lot under a given plan*

Principle of Indifference *each proposition of whose correctness we know nothing is endowed with a probability of* 1/2 *as is also its contradictory*

proper set inclusion *see subset*

residue *the remainder of an integer upon division by another integer called the modulus. Thus the residue of 19 modulo 7 is 5*

sampling with replacement *a sample is drawn from a set, examined, and then replaced so that the distribution in the population is the same as it was before. In sampling without replacement, since the sample remains outside, the distribution in the population is, in general, changed*

set inclusion *see subset*

significance level *a small probability chosen as the risk one is prepared to take that the rule of inference will lead to the rejection of an hypothesis that is true*

state-description *a state-description of x is a sentence which asserts for every attribute* C_j *predicable of x whether it applies or not*

statistical inference (a) *the making of inferences about the state of nature on the basis of incomplete data that involve randomness* (b) *direct statistical inference – from the characteristics of a population, conclusions are inferred about a sample drawn from that population,* (c) *inverse statistical inference – information about a sample is used to draw a conclusion about the population from which the sample is drawn;* (d) *predictive statistical inference – conclusions are drawn from a sample of a population about another sample not overlapping the first*

subset *a set B is a subset of A if every element of B is an element of A. The relation between A and B is that of set inclusion. If not every element of A is also an element of B, then we say B is a proper subset of A, and the relationship is one of proper set inclusion*

Appendix C

LIST OF SYMBOLS

$A(x)$ x has property A
$\neg A(x)$ not-$A(x)$; x has not property A
$P \rightarrow Q$ if P then Q; P yields Q
$x = y$ x equals y
$x \neq y$ x does not equal y
$x < y$ x is less than y; x precedes y
$x \leqq y$ x is less than or equal to y
$x > y$ x is greater than y; x follows y
$x \geqq y$ x is greater than or equal to y
$A \& B$ A and B
$x \in A$ x is an element of the set A; x belongs to A
$\{x, y, z\}$ the set consisting of the elements x, y, and z
$A \supset B$ the set A contains the set B as a subset
$A \supseteq B$ the set A contains the set B or is identical with B
$A \subset B$ the set A is contained in the set B as a subset
$A + B$ the union of the sets A and B, namely, the set which contains every element which belongs to either A or to B and no others
$A \cap B$ the intersection of the sets A and B, namely, the set which contains every element which belongs to both A and B and no others

$\lim_{n \to \infty} X_n$ the limit of X_n as n grows beyond all bounds

$|x|$ the absolute value of x, namely the larger of $-x$ and x

$a \cdot b$ a multiplied by b. This form is equivalent to the more common $a \times b$.

$n!$ n factorial. This represents the product of the first n integers. Thus $5! = 1.2.3.4.5$

$\binom{n}{k}$ the number of combinations of n things taken k at a time

$$\binom{n}{k} = \frac{n!}{(n-k)!k!}$$

$p(H)$ the probability of H

$p(H|E)$ the conditional probability of H given E

$a:b$ the ratio of a to b, namely a/b

λ the Greek letter λ, used to represent the *likelihood ratio*

Appendix D

ALPHABETICAL LIST OF TRACTATES IN THE MISHNAH

Arakhin
Avodah Zarah
Avot
Bava Batra
Bava Kama
Bava Metzia
Bekhorot
Berakhot
Betzah
Bikkurim
Demai
Eduyot
Eiruvin
Gittin
Horayot
Ḥagigah
Ḥallah
Ḥullin
Kelim
Keritot
Ketubot
Kiddushin
Kil'aim
Kinnim
Maasrot
Maaser Sheni
Makhshirin
Makkot
Megillah
Me'ilah
Menaḥot
Middot
Mikvaot
Mo'ed Katan
Nazir
Nedarim
Nega'im
Niddah
Ohalot
Orlah
Parah
Pe'ah
Pesaḥim
Rosh haShannah
Sanhedrin
Shabbat
Shekalim
Shevi'it
Shevu'ot
Sotah
Sukkah
Taanit
Tamid
Temurah
Terumot
Tevul Yom
Toharot
Uktzin
Yadaim
Yevamot
Yoma
Zavim
Zevaḥim

Appendix E

TALMUDIC TEACHERS (referred to in the text)

T or A following the name referred to in the text represent *Tanna* and *Amora* respectively.

ABAYE A 280–338?
RABBI ABBAHU A *fl. ca.* 275
RABBI BANA'AH T end of second century
BAR KAPPARA T end of second century
RABBI ELAZAR A died *ca.* 279
RABBI ELIEZER T *fl.* 90–130
RABBI ḤEZEKIAH A *fl. ca.* 350
RABBI ISAAC BEN ELAZAR A *fl. ca.* 250
RABBI ISHMAEL T *fl.* 90–115, redactor of *Mekhilta*
RABBI JONAH A *fl. ca.* 350
RABBI JUDAH T *fl.* 130–160
RABBI JUDAH THE PRINCE T 135–193?, editor of the Mishnah
RABBI MEIR T *fl.* 130–160
RABBI NAḤMAN A died 329
RABBI NEḤEMIAH T *fl.* 130–160. Solomon Gandz ascribes to R. Neḥemiah the redaction of the oldest extant Hebrew geometry, the *Mishnat haMiddot**
RABBI PEDAYAH A *fl.ca.* 300

RAV (ABBA ARIKHA) A died 247, founder of the college at Sura in 219

RAVA A 299–352

RABBI SAMUEL A *fl. ca.* 275, one of several contempary third generation Amoraim of the same name. Since our text omits Rabbi Samuel's patronymic, it is not clear which Samuel is meant.

RABBI SHAMMAI A *fl. ca.* 350

RABBAN SHIMON BEN GAMLIEL T *fl. ca.* 150

RABBI SHIMON BEN SHETAḤ T first century B.C.E.

RABBI SHIMON SHEZURI T *fl. ca.* 150

RABBI YOḤANAN A died *ca.* 279, reputed to have lived over 100 years, headed the college at Tiberias

RABBI YOSÉ T *fl. ca.* 175

RABBI YOSÉ BEN RABBI BON A the last Amora mentioned by name in the Jerusalem Talmud, second half of the fourth century

RABBI YOSÉ BEN KIPPAR T *fl. ca.* 175

RABBI Z'EIRA A several teachers by this name are known. Rabbi Z'eira I *fl. ca.* 275 and Rabbi Z'eira II lived half a century later

* Solomon Gandz, 'The Mishnat haMiddot ... ,' *Quellen und Studien zur Geschichte der Mathematik, Astronomie und Physik,* Abt. A, vol.2 (Berlin, 1932), 10

Bibliography

RABBINIC WORKS

Dates are given for all authors except recent and contemporary ones.

ABRAHAM BEN DAVID OF POSQUIERES (1125–1198) *Glosses on Maimonides Mishneh Torah,* printed in ed. (1) of *Mishneh Torah*

ABRAHAM KOHEN PIMINTEL (end of sixteenth century) *Minḥat Kohen* (Amsterdam, 1668)

ABRAMSKI, YEḤEZKEL *Ḥazon Yeḥezkel* (Commentary on Tosefta) vol. I (Warsaw, 1925)

ARAMAH, ISAAC BEN MOSHEH (1420–1494) *Akedat Yitzḥak* (5 vols.), ed. H.J. Pollack, Pressburg, 1849

ASHER BEN YEḤIEL (1250?–1328) *Code* printed in edition of the Babylonian Talmud

– *Responsa* (Vilna, 1885)

ASHKENAZI, BEZALEL (1520?–1589) *Shittah Mekubetzet* Bava Kama, Bava Metzia, Bava Batra (Warsaw, 1878); Betzah (Ofen, 1820); Ketubot (Lemberg, 1837)

BEN-ZIMRA, DAVID BEN SHLOMO (1479–1589) *Commentary on Maimonides' Mishneh Torah*, printed in edition (1) of Mishneh Torah

BIBLE with commentaries of Rashi, Ibn Ezra, Saadya Gaon, Gersonides, Kimhi, etc. (New York: Schoken, N.Y., 1946)

– Revised Standard Version (New York: Thomas Nelson & Sons, 1952)

CARO, JOSEPH (1488–1575) *Kessef Mishneh*, printed in edition (1) of Maimonides' *Mishneh Torah*

– *Shulhan Arukh Ḥoshen Mishpat*, photo-offset 1898 ed. (Jerusalem, 1959)

COLON, JOSEPH (1420?–1480) *Responsa* (1) Venice 1519, (2) Lemberg 1798

CORCOS, JOSEPH (*fl*. sixteenth century) *Commentary on Maimonides' Mishneh Torah* in *Sefer haHashlamah al Mishneh Torah* (New York, 1947)

CRESCAS, ḤASDAI (1340–1410) *Or haShem*, photo-offset of Vienna edition (Tel Aviv 5723, 1963)

DONNOLO, SHABBETAI BEN ABRAHAM (913–970) *Sefer Ḥakmoni* (Commentary on *Sefer Yetsirah*), printed in ed. of *Yetsirah*

DUNASH IBN TAMIM ABU SAHL (tenth century) *Commentary on Sefer Yetsirah* ed. M. Grossberg (London, 1902)

ENGEL, JOSEPH, *Gilyonei haShass*, printed in Jerusalem (1960) edition of Jerusalem Talmud

EVEN-SHMUEL, JUDAH *Maimonides' Moreh Nebukim with Commentary*, critical edition, 3 vols. covering Part I and Part II, chaps. 1–24 (Jerusalem, 1958)

FALK, JACOB JOSHUA (1680–1754) *Penei Yehoshua*, 3 vols. (New York, undated photo-offset)

FALK-KOHEN, JOSHUA (*d*. 1616) *Derishah U-Perishah*, printed in Jacob ben Asher, *Tur*, 7 vols. (New York, 1953)

GERSHOM BEN JUDAH (960–1040) *Commentary on the Babylonian Talmud*, printed in *Talmud* edition

GERSHON BEN SHLOMO (thirteenth century) *Shaar haShamayim* (Warsaw, 1875)

GERSONIDES, see Levi ben Gershon

IBN EZRA, ABRAHAM (*d*. 1167) *Commentary on Isaiah*, printed in Bible edition

– *Sefer ha-Mispar*, ed. M. Silverberg (Frankfurt a.M., 1895)

– *Ha-Olam* (1) transl. into English by Jekuthiel Ginsburg, "Rabbi ben Ezra on Permutations and Combinations," *The Mathematics Teacher* 15 (1922), pp. 347–56

(2) Latin transl. of 1281, in Marshall Clagett, *Nicole Oresme ...* (Madison, 1968), pp. 445–6

ISAAC BAR SHESHET (1326–1408) *Responsa* (Lemberg, 1805)

JACOB BEN ASHER (*d.* 1340) *Tur* with standard commentaries, 7 vols. (New York, 1953)

JEHUDAH BEN BARZILAI (twelfth century) *Commentary on Sefer Yetsirah*, ed. S.J. Halberstam with add. notes by D. Kaufmann (Berlin, 1885)

JOSEPH DI TRANI THE ELDER (sixteenth century) *Responsa* Part II (Furth, 1768)

LEVI BEN GERSHON (1288–1344?) *Commentary on Pentateuch,* 2 vols. undated photo-offset of Venice 1547 ed.

– *De Numeri harmonicis* in J. Carlebach, *Lewi ben Gerson als Mathematiker* (Berlin, 1910)

– *Milḥamot haShem,* undated photo-offset ed. (Riva, 1540)

– *Sefer Maasei Ḥoshev*, ed. G. Lange (Frankfurt a.M., 1909)

LIEBERMAN, SAUL *Tosefta Ki-Fshutah*, critical ed. of the *Tosefta*, with a comprehensive commentary, Part I (New York, 1955)

LURIA, SHLOMO BEN YEḤIEL (1510–1573) *Yam shel Shlomo*, 2 vols. photo-offset of Stettin 1861 ed. (New York, 1968)

MAIMONIDES, MOSES (1135–1206) *Commentary on the Mishnah*: (1) Hebrew transl. by Samuel ibn Tibbon *et al.*, printed in B. Talmud ed.; (2) Hebrew transl. and notes by Joseph Kapah, 3 vols. (Jerusalem, 1963–67)

– *Mishneh Torah*: (1) with glosses of R. Abraham of Posquieres and commentaries *Kessef Mishneh*, *Mishneh LaMelekh,* R. David ben Zimra, etc., 5 folio vols. (Wilna, 1900); (2) with a modern commentary, 16 vols. (Jerusalem, 1957–65); (3) *The Code of Maimonides transl. into English,* ed. Julian Oberman. Yale Judaica Series (Yale University Press, 1949 [incomplete])

– *Moreh Nevukhim*: (1) *Dalalat al-haïrin* (the Arabic text as established by S. Munk), ed. J. Junovitch (Jerusalem, 1931); (2) Hebrew transl. by Samuel ibn Tibbon and 4 commentaries, Warsaw 1872; (3) transl. Ibn Tibbon, critical ed. and commentary by Judah EvenShmuel (Jerusalem, 1959); (4) Latin transl. made early in thirteenth-century ed. Augustinus Justinianus, photo-offset Paris 1520 (Frankfurt a.M.: Minerva, 1964); (5) *Le Guide des Egarés,* French transl. by S. Munk, 3 vols. (Paris, 1856–66); (6) *The Guide for the Perplexed,* English transl. by M. Friedlander, 2nd ed. (London, 1936); (7) *The Guide for the Perplexed,* English transl. by Shlomo Pines with intro. by Leo Strauss (Chicago, 1963)

– *Sefer haMitzvot* (1) Hebrew transl. by Samuel ibn Tibbon and classical commentaries (Jerusalem, 1956); (2) Hebrew transl. by J. Kapah (Jerusalem, 1958); (3) English transl. by C.B. Chavel, 2 vols. (London, 1967)

– *Treatise on Logic – Original Arabic and 3 Hebrew translations,* ed. & transl. into English by I. Efros (New York, 1938)

– *Responsa,* 3 vols., ed J. Blau (Jerusalem, 1958–61)

MARGALIOT, MOSHEH (*d.* 1780) *Penei Mosheh* (Commentary on Jerusalem Talmud), printed in ed. of J. Talmud

MEIRI, MENAḤEM BEN SHLOMO (1249–1319?) *Beth haBeḥirah-Bava Batra*, ed. Abraham Schreiber (New York, 1956)

– *Beth haBeḥirah-Niddah*, ed. Abraham Schreiber (New York, 1949)

Mekhilta (first century) with *Malbim* Commentary (Wilna, 1891)

MISHNAH (second century) (1) with standard commentaries, 12 vols. photo-offset of Vilna ed. (New York, 1953); (2) with Maimonides' Commentary, ed. J. Kapaḥ, 3 vols. (Jerusalem, 1963–7); (3) with introduction and commentary by H. Albek, 7 vols. (Jerusalem, 1958–9); (4) ed. and transl. into English by Philip Blackman, 7 vols. (London, 1951–6). The Mishnah is also included as part of the text of both Talmuds

MORDECAI BEN HILLEL ASHKENAZI (*d.* 1298) *Mordecai,* printed in ed. of Babylonian Talmud

MOSHEH BEN NAḤMAN (1194–*ca.* 1270) *Novellae to Ḥullin*, ed. Solomon Reichman (New York, 1955)

NATHAN BEN YEḤIEL (1035?–1106) *Sefer ha-Aruch,* ed. Alexander Kohut, reprint (New York, 1955?)

NISSIM BEN REUVEN (1320–1380) *Commentary to Alfasi's Code,* printed in B. Talmud, ed.

– *Novellae to Sanhedrin* (New York, 1946)

Pesikta de-Rav Kahana (fourth century), ed. Bernard Mandelbaum (New York, 1962)

RASHI (1040–1105) *Commentary on Babylonian Talmud,* printed in B. Talmud ed.

RIDBAZ, JACOB DAVID (1845–1913) *Commentary on Jerusalem Talmud,* printed in ed. of J. Talmud

ROSANES, JUDAH (1657–1727) *Mishneh LaMelekh*: (Commentary on Maimonides' *Mishneh Torah*), printed in ed. (1) of *Mishneh Torah*

SAMUEL BEN MEIR (1085?–1174?) *Commentary on Babylonian Talmud*, printed in Talmud ed.

SHLOMO BEN ABRAHAM BEN ADRET (1235–1310) *Commentary on the Talmud* (Jerusalem, 1954)

– *1255 Responsa* (Vienna, 1812)

– *Torat Habayit ha-Arokh,* undated ed.

SHLOMO YITZḤAKI, see Rashi

Sifra (second century) with classical commentaries (Jerusalem 1959)

Sifri (first and second centuries) with *Malbim* commentary (Wilna, 1891)

DI TRANI, see Joseph di Trani the Elder

TALMUD (a) *Babylonian Talmud* (1) with standard commentaries, *Rashi, R. Samuel ben Meir, Tosafot* etc.; Maimonides' *Commentary on the Mishnah*; codes of *R. Asher ben Yeḥiel, Alfasi, Mordecai*; and *Tosefta,* 20 vols. (New York, 1948); (2) English transl. ed. I. Epstein, 18 vols. (London, 1961). (b) *Jerusalem Talmud* (1) with standard commentaries, *Penei Mosheh*, *Ridbaz,* etc., 7 vols. (New York, 1959); (2) with standard commentaries and Rabbi Joseph Engel's *Gilyonei haShass* etc., 8 vols. (Jerusalem, 1960)

TOSAFOT (thirteenth century) Notes on B. Talmud, printed in Talmud ed.

TOSEFTA (second century) (1) with classical commentaries in B. Talmud ed.; (2) with commentary by Yeḥezkel Abramski, part I (Warsaw, 1925); (3) critical ed. with commentary by Saul Lieberman, part I (New York, 1955)

YEDAIAH BEN ABRAHAM BEDERSI (1270–1340) 'Letter to Rabbi Shlomo ben Adret,' *1255 Responsa of Rabbi Shlomo ben Adret*, no. 418

YESHUAH HALEVI BEN JOSEPH (1440–1500) *Halikhot Olam*: with commentaries by Rabbi Joseph Caro and Rabbi Shlomo Algazi (Jerusalem, 1960)

Yetsirah (1) *Das Buch der Schopfung* (*Jesirah*): critical ed. with introduction and notes and German transl. by L. Goldschmidt (Frankfurt a.M., 1894); (2) with classical commentaries including Rabbi Shabbetai Donnolo (Jerusalem, 1962)

YOM TOV BEN ABRAHAM (fourteenth century) *Commentary on the Talmud*, 6 vols. (Tel Aviv, 1958)

GENERAL WORKS

ALBRIGHT, W.F. *The Archaeology of Palestine* (Pelican, 1960)

AQUINAS, THOMAS (1225?–1274) *Exposition of the Posterior Analytics of Aristotle*, translated into English by Pierre Conway (Quebec: Librairie Phil. M. Doyon, 1956)

– Summa Theologiae, 5 vols. (Ottawa: Dominican College, 1941–5)

ARCHIBALD, RAYMOND C. 'Babylonian mathematics' *Isis* 26 (1936), 63–81

ARISTOTLE *The Works of Aristotle*, transl. into English, ed. W.D. Ross, 12 vols. (Oxford, 1930–52)

BARKER, STEPHEN F. 'The role of simplicity in explanation,' *Current Issues in the Philosophy of Science*, ed. Feigl, H. & Maxwell, G., 1959

BAYES, THOMAS 'An essay towards solving a problem in the doctrine of chances,' *Philosophical Transactions* (Royal Society of London), vol. 53 (1793), 370–418

BENNETT, JONATHAN 'Some aspects of probability and induction *I*,' *British Journal for Philosophy of Science*, 7 (1956–7), 220–30

BERNOULLI, JACOB *Ars Conjectandi* (Basle, 1713), photo-offset (Brussels, 1968)

BLACK, MAX 'Probability,' *The Encyclopedia of Philosophy* (New York: Macmillan, 1967), vol. VI, 464–79

BLOM, SIRI 'Concerning a controversy on the meaning of probability,' *Theoria*, 21 (1955), 65–98
BOREL, EMILE 'Apropos of a treatise on probability,' reprinted in *Studies in Subjective Probability*: ed. Kyburg and Smokler, 1964
– *Probability and Certainty*, transl. into English by Douglas Scott (New York, 1963)
BOYER, C.B. 'Fundamental steps in the development of numeration,' *Isis* 35 (1944), 153–68
BOYER, CARL B. *The History of the Calculus and its Conceptual Development* (New York: Dover, 1949)
– *A History of Mathematics* (New York, 1968)
BRAITHWAITE, RICHARD BEVAN *Scientific Explanation* (Cambridge, 1964)
BYRNE, EDMUND F. *Probability and Opinion* (The Hague, 1968)
CALLUS, D.A., ed. *Robert Grosseteste Scholar and Bishop* (Oxford, 1955)
CAMERON, GEORGE G. 'The Babylonian scientist and his Hebrew colleague,' *Biblical Archaeologist* 8 (1944) 21–9, 32–40
CANTOR, MORITZ *Vorlesungen uber Geschichte der Mathematik*, 2nd ed., 4 vols. (Leipzig, 1894)
CARLEBACH, J. *Lewi ben Gerson als Mathematiker* (Berlin, 1910)
CARNAP, RUDOLF *Logical Foundations of Probability,* 2nd ed. (Chicago, 1962)
– 'Remarks on probability,' *Philosophical Studies* 14 (1963) 65–75
– 'The aim of inductive logic,' *Proc. 1960 International Congress – Logic, Methodology and Philosophy of Science,* p. 303–18
– 'The two concepts of probability,' reprinted in *Readings in the Philosophy of Science,* ed. Feigl and Broadbeck (1953), pp. 438–55
CHERNOFF, HERMAN, and MOSES, LINCOLN *Elementary Decision Theory* (New York: Wiley, 1959)
CLAGETT, MARSHALL *Nicole Oresme and the Medieval Geometry of Qualities and Motions* (University of Wisconsin Press, 1968)
COUTURAT, LOUIS *La Logique de Leibniz d'après des documents inédits* (Paris, 1901)

CROMBIE, A.C. *Augustine to Galileo*, 2 vols (London: Mercury Books, 1961)

CURTZE, MAXIMILIAN 'Die Abhandlung des Levi ben Gerson über Trigonometrie und dem Jacobstab,' *Bibliotheca Mathematica* N.S. 12 (1898), 97–112

CZERWINSKI, ZBIGNIEW 'On the relation of statistical inference to traditional induction and deduction,' *Studia Logica* VII (1958), 243–60

DAUBE, DAVID 'Rabbinic methods of interpretation and Hellenistic rhetoric.' *Hebrew Union College Annual* 22 (1949), 239–64

DAVID, F.N. *Games, Gods and Gambling* (New York, 1962)

DAY, JOHN P. *Inductive Probability* (New York, 1961)

DEMORGAN, AUGUSTUS *An Essay on Probabilities* (London, 1838)

DERRY, T.K., and WILLIAM, T.I. *A Short History of Technology* (Oxford, 1960)

EFROS, ISRAEL *Philosophical Terms in the Moreh Nebukim* (New York: Columbia University Press, 1924)

EHRENFELD, SYLVAIN, and LITTAUER, S.B. *Introduction to Statistical Method* (New York, 1964)

ELLIS, R.L. 'On the foundations of the theory of probabilities,' *Transactions of the Cambridge Philosophical Society*, vol. 8, no. 1 (1849) pp. 1–6

FEIGL, HERBERT, and BRODBECK, MAY *Readings in the Philosophy of Science* (New York, 1953)

FERGUSON, THOMAS S. *Mathematical Statistics* (New York, 1967)

FEYERABEND, PAUL K. 'Comments on Baker's *The Role of Simplicity in Explanation*,' *Current Issues in the Philosophy of Science* (1959), 278–80

DE FINETTI, BRUNO 'Foresight: Its logical laws, its subjective sources,' reprinted in *Studies in Subjective Probability*, ed. Kyburg and Smokler, 1964

FINNEY, D.J. *Statistics for Mathematicians* (Edinburgh: Oliver & Boyd, 1968)

FISHER, R.A. *Contributions to Mathematical Statistics* (New York, 1950)

– *Statistical Methods and Scientific Inference* (London, 1956)

GANDZ, SOLOMON 'On the origin of the term root,' *American Mathematical Monthly* 33 (1926), 261–5

– 'The origin of the Ghubar numerals or the Arabian abacus and the articuli,' *Isis* 16 (1931), 393–424

– 'The origin of the gnomon,' *Proc. American Academy for Jewish Research* 2 (1931), 23–38

– *The Mishnat Ha-Middot*, the first Hebrew Geometry about 150 C.E., and the *Geometry of al Khowarizmi*, the first Arabic geometry (*ca.* 820) representing the Arabic version of the *Mishnat ha-Middot. Quellen und Studien ur Geschichte der Mathematik Astronomie und Physik*, Abt. A vol. 2. (Berlin: Springer, 1932)

– '*Mene, Mene, Tekel, Upharsin*. A chapter in Babylonian mathematics,' *Isis* 26 (1936), 82–94

– 'Origin and development of quadratic equations in Babylonian, Greek and early Arabic algebra,' *Osiris* 3 (1937), 405–558

– 'The algebra of inheritance,' *Osiris* 5 (1938), 319–91

– 'Saadya Gaon as a mathematician,' *American Academy for Jewish Research Texts and Studies* II, Saadya vol. (1943), 141–95

GARDEIL, A. 'La "Certitude Probable,"' *Revue des Sciences Philosophiques et Théologiques* 5 (1911), 237–66, 441–85

– 'La Topicité,' *Revue des Sciences Philosophiques et Théologiques* 5 (1911), 750–7

GILSON, ETIENNE *Le Thomisme* (Paris, 1923)

GINSBURG, JEKUTHIEL 'Rabbi ben Ezra on permutations and combinations,' *The Mathematics Teacher* 15 (1922), 347–56

GOETSCH, H. 'Die Algebra der Babylonier,' *Archive for History of Exact Sciences* 5 (1968), 79–153

GOLDSTEIN, BERNARD R. 'Preliminary remarks on Levi ben-Gerson's contributions to astronomy,' *Proc. of the Israel Academy of Sciences and Humanities*, vol. 3 (1969), 239–54

– 'Theory and observation in medieval astronomy,' *Isis* 63 (1972), 39–47

GOUDGE, THOMAS A. *The Thought of C.S. Pierce* (Toronto, 1950)

GRAETZ, HEINRICH *History of the Jews*, 6 vols. (Philadelphia, 1946)

GRANT, EDWARD 'Nicole Oresme and the commensurability or incommensurability of the celestial motions,' *Archive for History of Exact Sciences* I (1961), 420–58

– *Nicole Oresme: De proportionibus proportionum and Ad pauca respicientes*, ed. and transl. into English (University of Wisconsin Press, 1966)

GUGGENHEIMER, HEINRICH 'Ueber ein Bemerkenswertes Logisches System aus der Antike,' *Methods* (1951), 150–64

– 'Logical problems in Jewish tradition,' in *Confrontations with Judaism*, ed. Philip Longworth (London: Anthony Blond Ltd., 1966), 171–96

– 'Die Dialektische Philosophie im Talmud,' *Proc. XI International Congress Philosophy,* Brussels, 1953

HANSON, N.R. 'Is there a logic of scientific discovery?' *Australasian Journal of Philosophy* 38 (1960), 91–106

– *Patterns of Discovery* (Cambridge University Press, 1965)

– 'The idea of a logic of discovery,' '*Dialogue* 4 (1965–66), 49–61

HASOFER, A.M. 'Random mechanisms in Talmudic literature,' *Biometrika* 54 (1966), 316–21

– 'Some aspects of Talmudic probabilistic thought,' *Proc. Association of Orthodox Jewish Scientists* 2 (1969), 63–80

HEMPEL, CARL G. *Deductive-Nomological vs. Statistical Explanation,* Minnesota Studies in the Philosophy of Science, vol. 3., ed. Feigl and Maxwell (Minneapolis, 1962)

HILPINEN, RISTO *Rules of Acceptance and Inductive Logic* (Amsterdam, 1968)

HINTIKKA, J., and HILPINEN, R. 'Knowledge, acceptance and inductive logic,' *Aspects of Inductive Logic,* ed. J. Hintikka and P. Suppes (Amsterdam, 1966), pp. 1–20

HOEL, PAUL G. *Introduction to Mathematical Statistics,* 3rd ed. (New York, 1962)

HUME, DAVID *An Enquiry Concerning Human Understanding*, ed. C.W. Eliot (New York, 1910)

– *A Treatise of Human Nature*, 2 vols., ed. T.H. Green and T.H. Grose (London, 1874)

HUSIK, ISAAC *A History of Mediaeval Jewish Philosophy* (Philadelphia, 1946)

JAMMER, MAX *Concepts of Space* (Harvard University Press, 1954)

JEFFREYS, HAROLD *Theory of Probability* (Oxford, 1961)
KAPP, REGINALD O. 'Ockam's Razor and the unification of physical science,' *British Journal for the Philosophy of Science* 8 (1958), 265–80
KATZ, JERROLD J. *The Problem of Induction and Its Solution* (University of Chicago Press, 1962)
KEMENY, JOHN G. 'The use of simplicity in induction,' *The Philosophical Review* 62 (1953), 391–408
KEYNES, JOHN MAYNARD *A Treatise on Probability*, 3rd ed. (Oxford, 1961)
KING, AMY C., and READ, CECIL B. *Pathways to Probability Theory. History of the Mathematics of Certainty and Chance* (New York, 1963)
KNEALE, WILLIAM *Probability and Induction* (Oxford, 1949)
KOLMOGOROV, A.N. *Foundations of the Theory of Probability, English transl. by Nathan Morrison from German original, 1933 (Chelsea, 1950)*
KOOPMAN, BERNARD O. 'The bases of probability,' reprinted in *Studies in Subjective Probability,* ed. Kyburg & Smokler (New York, 1964)
KYBURG, HENRY E., JR. *Probability and the Logic of Rational Belief* (Middletown: Wesleyan University Press 1961)
– 'A further note on rationality and consistency,' *Journal of Philosophy,* 60 (1963), 463–5
– *Probability and Inductive Logic* (London; Macmillan, 1970)
KYBURG, HENRY E., and SMOKLER, HOWARD E. *Studies in Subjective Probability* (New York: Wiley, 1964)
LANGE, G. *Sefer Maasei Choscheb, Die Praxis des Rechners, ein hebraisch – arithmetisches Werk des Levi ben Gerschom aus dem Jahre 1321* (Frankfurt a.M., 1909)
LEBLANC, HUGUES *Statistical and Inductive Probabilities* (New Jersey: Prentice Hall, 1962)
LEFF, GORDON *Gregory of Rimini* (Manchester University Press, 1961)
LEHRER, KEITH 'Induction, reason and consistency,' *British Journal for Philosophy of Science* 21 (1970), 103–14
– 'Knowledge and probability,' *The Journal of Philosophy* 61 (1964), 368–72

LEVI, ISAAC *Gambling with Truth* (New York: Knopf, 1967)
LIEBERMAN, SAUL *Hellenism in Jewish Palestine* (New York, 1950)
LINDBERG, D.C. 'The theory of pinhole images in the fourteenth century,' *Archive for History of Exact Sciences* 6 (1970), 299–325
LINDGREN, B.W. *Statistical Theory* (Macmillan, 1968)
LINDGREN, B.W. and MCELRATH, G.W. *Introduction to Probability and Statistics,* 2nd ed. (Macmillan, 1966)
LOEWY, ALFRED 'Uber die Zahlbezeichnung in der jüdischen Literatur,' *Jeschurun* 17, 202–12
MAHALANOBIS, P.C. 'The foundations of statistics,' *Sankya* 18 (1957), 183–94
MAIER, ANNELIESE *An der Grenze von Scholastik und Naturwissenschaft* (Essen, 1943)
– *Zwei Grundprobleme der Scholastischen Naturphilosophie* (Rome, 1951)
MENNINGER, KARL *Zahlwort und Ziffer,* 2 vols. in 1 (Gottingen, 1958)
MICHELL, JOHN 'An inquiry into the probable parallax and magnitude of the fixed Stars, etc.,' *The Philosophical Transactions of the Royal Society of London*, vol. 57, XXVII, 234. Abridged ed., vol. 12 (1809), 423–40
MILL, JOHN STUART *A System of Logic*, 8th ed. (New York, 1888)
VON MISES, RICHARD *Probability, Statistics and Truth* (London: Hodge, 1939)
MOIVRE, ABRAHAM DE *The Doctrine of Chances*, reproduction of 2nd ed., 1738 (London, 1967)
MUNTNER, SUESSMAN *R. Shabtai Donnolo* (Jerusalem, 1949)
NAGEL, Ernest 'Principles of the theory of probability,' *International Encyclopedia of Unified Science* I, no. 6 (1939)
NEUGEBAUER, OTTO 'The astronomy of Maimonides and its sources,' *Hebrew Union College Annual* 22 (1949), 321–63
– 'The transmission of planetary theories in ancient and medieval astronomy,' *Scripta Mathematica* 22 (1956), 165–92
– *The Exact Sciences in Antiquity* (Providence: Brown University Press, 1957)

O'CONNOR, D.J. 'Determinism and predictability,' *British Journal for the Philosophy of Science* 7 (1956–59), 310–15

O'DONNELL, J. REGINALD 'Nicholas of Autrecourt,' *Medieval Studies,* 1 (1939)

– 'The philosophy of Nicholas of Autrecourt and his appraisal of Aristotle,' *Medieval Studies*, 4 (1942), 97–125

O'DONNELL, TERENCE, *History of Life Insurance* (Chicago, 1936)

ORE, OYSTEIN, 'Pascal and the invention of probability theory,' *American Mathematical Monthly* 67 (1960), 409–19

ORESME, NICOLE *De proportionibus proportionum* and *Ad pauca respicientes*, ed. & transl. into English by Edward Grant (University of Wisconsin Press, 1966)

– *Tractatus de configurationibus qualitatum et motuum,* ed. & transl. into English by Marshall Clagett (University of Wisconsin Press, 1968)

PAP, ARTHUR *An Introduction to the Philosophy of Science* (New York, 1962)

PASCAL, BLAISE *Œuvres de Blaise Pascal*, vol. IV, ed. Henri Massis (Paris, 1927)

PIERCE, CHARLES S. *Collected Papers*, ed. Charles Hartshorne and Paul Weiss (Harvard University Press, 1931–58)

PERELMAN, CHAIM, and L. OLBRECHTS-TYTECA *The New Rhetoric*, transl. by j. Wilkinson and P. Weaver (University of Notre Dame Press, 1969)

PHILO (with an English translation by F.H. Cohen and G.H. Whitaker), vol. 1, *De Opificio Mundi* ... (London, 1929)

PLATO *The Dialogues of Plato*, 5 vols., transl. into English by B. Jowett (Oxford, 1875)

– *The Works of Plato*, vol. V *The Laws*, transl. into English by George Burges (London, 1896)

POPPER, KARL R. 'The Propensity interpretation of probability,' *British Journal for the Philosophy of Science* 10 (1959), 25–42

– *The Logic of Scientific Discovery* (New York: Harper, 1968)

RABINOVITCH, NACHUM L. 'Probability in the Talmud,' *Biometrika* 56 (1969), 437–41

– 'Combinations and probability in Rabbinic literature,' *Biometrika* 57 (1970), 203–5

– 'Rabbi Levi Ben Gershon and the origins of mathematical induction,' *Archive for History of Exact Sciences* 6 (1970) 237–48

RAMSEY, FRANK PLUMPTON 'Truth and Probability,' reprinted in *Studies in Subjective Probability*, ed. Kyburg and Smokler (1964)

REICHENBACH, HANS 'The logical foundations of the concept of probability,' transl. and reprinted in *Readings in the Philosophy of Science*, ed. Feigl and Brodbeck (1953)

– *The Theory of Probability* (University of California Press, 1949)

RESCHER, NICHOLAS 'On prediction and explanation' *British Journal for the Philosophy of Science* 8 (1958) 281–90

– ed. *The Logic of Decision and Action* (University of Pittsburgh Press, 1966)

– 'Semantic foundations for the logic of preference,' *The Logic of Decision and Action*, ed. N. Rescher (University of Pittsburgh Press, 1966), pp. 37–79

RUSSELL, BERTRAND *Human Knowledge: Its Scope and Limits* (London, 1948)

SAIDAN, A.S. 'Review of Edmund F. Byrne *Probability and Opinion*,' *Isis* 60 (1969), 255–7

SAMBURSKY, S. 'Possible and probable in ancient Greece,' *Osiris* 12 (1956), 35–48

SARFATTI, GAD BENAMI *Mathematical Terminology in Hebrew Scientific Literature of the Middle Ages* (Hebrew) (Jerusalem, 1968)

SARTON, GEORGE *Introduction to the History of Science*, 5 vols. (Baltimore, 1927–48)

SAVAGE, LEONARD J. 'The foundations of statistics reconsidered,' reprinted in *Studies in Subjective Probability*, ed. Kyburg and Smokler (1964)

SCHEFFLER, ISRAEL 'Explanation, prediction and abstraction,' *British Journal for the Philosophy of Science* 7 (1956–57), 293–309

SCHICK, F. 'Consistency and rationality,' *Journal of Philosophy* 60 (1963), 5–19

SCHWARZ, ADOLF *Der Hermeneutische Syllogismus in der Talmudischen Literature*, VII Jahresbericht der Israelitisch-Theologischen Lehranstalt in Wien, (Vienna, 1901)

– *Die Hermeneutische Analogie in der Talmudischen Literatur*, IV Jahresbericht der Israelitisch-Theologischen Lehranstalt in Wien (Vienna, 1897)
– *Die Hermeneutische Antinomie in der Talmudischen Literatur*, XX Jahnesbericht der Israelitisch-Theolischen Lehranstalt in Wien (Vienna, 1913)
– *Die Hermeneutische Induktion in der Talmudischen Literatur*, XVI Jahresbericht der Israelitisch-Theolischen Lehranstalt in Wien (Vienna, 1909)
– *Die Hermeneutische Quantitats relation in der Talmudischen Literatur*, XXIII Jahresbericht der Israelitisch-Theologischen Lehranstalt in Wien (Vienna, 1916)
SHEYNIN, O.B. 'Newton and the Classical Theory of Probability,' *Archive for History of Exact Sciences* 7 (1971) 217–43
SMITH D.E. *History of Mathematics*, 2 vols. (New York: Dover, 1958, reprint of 1953 ed.)
STEINSCHNEIDER, MORITZ *Mathematik bei den Juden*, ed. with index by Adeline Goldberg (Hildesheim, 1964)
STRACK, HERMANN L. *Introduction to the Talmud and Midrash* (Philadelphia, 1945)
SZECHTMAN, JOSHUA 'Notes on Dr Zukermann's introduction to his mathematical concepts in the Talmud,' *Scripta Mathematica* 25 (1960), 49–62
TODHUNTER, ISAAC *History of the Mathematical Theory of Probability* (Cambridge, 1865)
TOULMIN, STEPHEN E. *The Uses of Argument* (Cambridge, 1958)
TOZER, JOHN 'On the measure of the force of testimony in cases of legal evidence,' *Transactions of the Cambridge Philosophical Society*, vol. 8 (1843), 143–58
TURETSKY, MORRIS 'Permutations in the 16th century Cabala,' *Math. Teacher* 16 (1923), 29–34
VENN, JOHN *The Logic of Chance*, 3rd ed. (London, 1888)
WEINBERG, JULIUS R. *Nicolaus of Autrecourt* (Princeton, 1948)
– *Abstraction, Relation and Induction* (University of Wisconsin Press, 1965)
– *A Short History of Medieval Philosophy* (Princeton, 1964)
WEISSMANN, A. 'Remarks on the foundation of logic,' *Actes*

Congrès Internationale d'Histoire des Sciences VII (1953), 629–32

WORKMAN, ROLLIN W. 'Two extralogical uses of the Principle of Induction,' *Philosophical Studies* 13 (1962), 27–32

ZILSEL, E. 'The genesis of the concept of physical law,' *Philosophical Review* 51 (1942), 245–79

LEXICONS

JASTROW, MARCUS *A Dictionary of the Targumim, the Talmud, etc.*, 2 vols. (New York, 1950)

KOHUT, ALEXANDER *Aruch Completum*, 9 vols. (New York, 1955)

KLATZKIN, JACOB *Thesaurus Philosophicus Linguae Hebraicae*, 4 vols. (New York, 1968)

Index

Abraham ben David 5, 52, 70
acceptance number 85, 86
acceptance rules 36 ff., 44, 47, 61, 86, 98, 107, 118, 119, 130, 178
– arithmetic for 48–52
– incomplete definition of 45
– paradox of 98, 101–2, 105
Albertus Magnus 155
Al-Khuwarizmi 165
alternatives: counting 64 ff., 159; exclusive and exhaustive 59, 100, 102; logical 62 ff., 171; equiprobable xii, 72 ff., 89, 94, 102, 109, 127, 158, 160, 171; simple and compound 71; ultimate 64, 69
analogy 12 ff., 152
Aquinas, Thomas 74, 117, 155, 156, 168
Aramah, Isaac 27, 35, 92, 93
Aristotle 15, 16, 25, 31, 75, 77, 91, 122, 123, 124, 125, 126, 137, 152, 153, 154, 155, 156, 158, 159, 160, 166, 167, 168
Asher ben Yeḥiel 5, 11, 116, 127, 128, 170
Ashkenazi, Bezalel 6, 41, 54, 55
Axioms of Probability 56–7, 61, 134, 172
– axiom of multiplication proved 63

Bayes, Thomas 58, 164
Bayes's Theorem 58 ff., 116
Ben-Zimra, David 70
Bernoulli, Jacob 34, 63
Bernoulli, John 117
Boetius 143, 161
Bertrand's Paradox 99, 100
Blom, Siri 155
Braithwaite, R.B. 57
Byrne, E.F. 156, 168

Cardan 165

Carnap, Rudolf 17, 79, 80, 88, 107–109, 112
Caro, Joseph 6, 52
certainty: degrees of 9, 10, 42, 43, 111, 112, 123, 157, 171; lack of 18, 20, 153, 155, 171; like a 106, 116; science and 141, 152, 166
Chance, Laws of 64, 159
– random mechanisms 21–35
Clagett, Marshall 145, 148, 149, 150, 168
Colon, Joseph 6, 115, 116, 157
combinations and permutations xiii, 63, 66 ff., 76, 78, 142–51, 160, 161
confirmation 108 ff.
Corcos, Joseph 52
Crescas, Ḥasdai 77
Crombie, A.C. 166

David, F.N. 21, 88
decision 8, 10, 20, 65, 107, 109, 119, 128, 140
– procedure 48, 118, 173
– rule 130, 163, 178
– theory xii, 11, 172
– to believe 139
deduction: distinguished from induction 8, 11, 103, 104, 125, 152
Design: *vs* chance 26, 31, 123, 124
De Moivre, Abraham 62
Distribution 34, 55, 56, 62, 71, 73, 94, 124, 169
– hypergeometric 79
– in a sample 80–3, 101
– mean value of 79
Donolo, Shabbetai 143
double doubt 50, 54 ff., 71, 72

Ellis, R.L. 47
Engel, Joseph 52
equiprobable xii, 32, 63 ff., 69, 71 ff., 102, 128, 157–60
– a 'fair' game 21, 34
– chance 27, 35
– same relative frequency 25
– when are alternatives equiprobable? 76, 160
evidence xi, 108–17, 118, 135 ff., 141, 171
expectation 163, 164

Falk, Jacob Joshua 54
Falk-Kohen, Joshua 70
Fermat xi, 165
Finney, D.J. 37
Fisher, R.S. 124, 129
Frequency: empirical 55, 56, 73, 89; interpretation xii, 36–47, 57, 61, 62, 75, 103, 171, 172; relative 25, 26, 42, 45, 47, 48–52, 71, 82, 88, 98, 108, 109, 149

Gandz Solomon 143, 144, 161, 165, 166
Gardeil, A. 155, 156, 168
Gershom ben Judah 4, 116
Gershon ben Shlomo 122

Gilson, Etienne 74, 138, 166
Ginsburg, J. 145
Goldstein, B.R. 126, 151
Grant, E. 140, 149, 150, 151, 168
Guggenheimer, H. 17

Hasofer, A.M. xii, 33
Hempel, C.G. 110
Hilpinen, R. 102
Hintikka, J. 102
Hume, David 8
hypothesis testing xii, 37, 38, 118–30
– falsification 125 ff.

Ibn Ezra, Abraham 5, 145, 146, 157, 162, 163
Ibn Sina 74
Ibn Tibbon 157
indifference, principle of 63 ff., 73, 150, 178; paradox of, 96–7, 100–1
induction: and hypothesis 118 ff.; is ampliative 9; Mill's canons of 14; problem of 8, 89, 112, 113; rabbinic rules of 11 ff.
inductive inference 103
– direct 79, 80, 84
– inverse 80, 84, 88
– predictive 80, 88–91
Isaac bar Sheshet 5, 72–3

Jammer, Max 124
Jeffreys, Harold 163
Joseph di Trani 6, 65
Keynes, J.M. 9, 55, 63, 73, 96, 99, 124, 128–9, 163
Kneale, W. 43, 135, 141
Kolmogorov, A.N. 57
Kyburg, Henry E. 57, 100, 101, 102, 105, 134

Laplace 62
Law of Large Numbers 32, 34, 93, 178
Laws of Nature 73, 77, 113, 159,
– statistical 9, 91–3
Lehrer, Keith 101
Leibniz 117, 128–9, 163
Levi ben Gershon 5, 77, 126, 139, 140, 141, 147, 148, 150, 151, 158, 168, 169
Levi, Isaac 101
Lieberman S. 65
Likelihood: principle of maximum 127 ff.; ratio 130, 132
Lindgren, B.W. 79, 85, 103, 129
Locke 155
Luria, Shlomo 6, 20

Maimonides 1, 5, 6, 31, 32, 35, 41, 52, 67–70, 74, 75, 85, 90, 91, 103, 111, 112, 113, 114, 119, 122, 123, 135, 137, 139, 141, 150, 157, 164, 166, 167
Majority: counted and uncounted 38, 39, 42, 59, 106; follow the 36 ff., 55, 65, 80, 98, 103, 104, 112, 131, 152, 154, 159; higher 41, 43; majority of a 106; not due to chance 25; standard 43 ff., 59, 116

McElrath, G.W. 79, 85, 103, 129
Meiri, Menahem 5, 122
Michell, John 124
Mill, J.S. 9, 14, 113
Mordecai ben Hillel 5, 77
Mosheh ben Maimon *see* Maimonides
Mosheh ben Naḥman 5, 106

Newton, Isaac 124
Nicholas of Autrecourt 167, 168
Nissim ben Reuven 45, 157
Non-sufficient reason *see* indifference

Ockham's razor 121
O'Donnell, J.R. 167
Olbrechts-Tyteca, L. 129
operating characteristic 85–7, 178
Ore, Oystein 165
Oresme 140, 145, 148, 149, 150, 151, 168

Pacioli, Luca 165
Pap, Arthur 13
paradox 94–107
– lottery 101, 105
– of indifference 63, 96 ff.
particularisation, principle of 75, 122, 123
Pascal, Blaise xi, 163, 165
Perelman, C. 129
Philo 113
Pierce, C.S. 9, 75
Plato 144, 155
Popper, K.R. 45, 64, 75, 125
presumptions, legal 39, 42, 112–16
probabilism 166 ff.
probability: and opinion 153 ff.; classical definition 62, 75, 91; conditional 51 ff., 178; frequency interpretation *see* frequency; independent 53 ff., 178; inverse 58 ff.; posterior 60; prior 58 ff., 116, 119, 122; propensity interpretation 64, 74; subjectivist 133 ff.
probability function 56 ff., 61, 142
– hypergeometric 79
– range on [0, 1] 44, 57

Ramsay, F.P. 133
Rashi 4, 32, 33, 42, 60, 83, 99, 157, 171
reference class 44–7, 53, 64, 78
Reichenbach, Hans 57
Rescher, N. 72, 164
Ridbaz, Jacob David 95
Rosanes, Judah 70
Russell, Bertrand 57

Saadya Gaon 143, 144, 161
Sambursky, S. 160, 161
sampling 34, 78–91, 116, 179
Samuel ben Meir 76
Sarton, G. 143
Schwarz, A. 19
Sheynin, O.B. 124

Shlomo ben Adret 5, 40, 41, 71, 91, 106, 132
Shlomo ben Meir 44, 170
significance level 37, 41, 47, 130, 179
– test 122
Smith, D.E. 151, 161, 165
Socrates 155
statistical inference, use in induction 9
Strack, Herman, L. 6

Tartaglia 151, 165
Teleology 73, 124
Todhunter, Isaac xi
Toulmin, S. 172
Turetzky, M. 145

undecidable propositions 20, 137

Variability, in samples 30, 80, 82, 83
Venn, John 47, 142
von Mises, Richard 57, 96

Weinberg, J.R. 168

Xenokrates 160

Yedaiah ben Abraham Bedersi 90
Yeshuah Halevi ben Joseph 6, 14, 17
Yom Tov ben Abraham 5, 44, 106

Zilsel, E. 113

www.ingramcontent.com/pod-product-compliance
Ingram Content Group UK Ltd.
Pitfield, Milton Keynes, MK11 3LW, UK
UKHW032122310125
454513UK00004B/147